SpringerBriefs in Applied Sciences and Technology

Series editor

Andreas Öchsner, Southport Queensland, Australia

For further volumes:
http://www.springer.com/series/8884

Rita Ehrig · Frank Behrendt · Manfred Wörgetter
Christoph Strasser

Economics and Price Risks in International Pellet Supply Chains

Springer

Rita Ehrig
Christoph Strasser
Unit Resources and Technical Logistics
BIOENERGY 2020+ GmbH
Wieselburg-Land
Austria

Manfred Wörgetter
BIOENERGY 2020+ GmbH
Wieselburg-Land
Austria

Frank Behrendt
Chair Energy Process Engineering and Conversion Technologies for Renewable Energies
Berlin Institute of Technology (TU Berlin)
Berlin
Germany

ISSN 2191-530X ISSN 2191-5318 (electronic)
ISBN 978-3-319-07015-5 ISBN 978-3-319-07016-2 (eBook)
DOI 10.1007/978-3-319-07016-2
Springer Cham Heidelberg New York Dordrecht London

Library of Congress Control Number: 2014939044

Printed on acid-free paper

Springer is part of Springer Science+Business Media (www.springer.com)

Contents

Abstract

Purpose
This work investigates critical economic aspects and price risks along international pellet supply chains. This allows an estimation of risk margins in pellet trade and gives insight into crucial mechanisms, which drive pellet prices and worldwide trade.

Methodology/Approach
Supply costs for three real case studies are assessed with Canada, Australia and Russia as exporting countries and the EU as target market. Based on these, most significant economics and price indicators along the supply chain are identified and analysed. With these, the impact of several risks like raw material prices, exchange and freight rates on total prices is investigated.

Findings
Coincidently occurring price fluctuations within the supply chain can effect a 34–57 % variation of import prices. So, exchange rate volatility with more than 30 % variation between 2008 and 2011 has strongly hit individual pellet exporters to the EU. Nevertheless, the pellet price bears lesser risk than hard coal prices.

Research and Practical Implications
The assessment of various price data along the supply chain as well as interviews with pellets market actors allow to conclude how the pellet supply chain can be de-risked and how price risks are hedged to avoid project defaults and achieve consumers' renewable energy targets.

Originality/Value
A comprehensive review and analysis of pellet price risks has not been accomplished before and thus this work allows new insight into the interconnections between the sector, the various supply risks on the market and related de-risk strategies.

Chapter 1
Introduction

Abstract This chapter outlines the role of biomass imports to the EU and thus the importance of reliable supply to the consumers. Biomass plays a major role to fulfil the EU's energy targets for 2020. For reaching the ambitious energy targets, the EU member states will rely on biomass imports from non-EU countries. Economics and reliability of pellet supply are key issues in international trade. Pellet economics and pricing is characterised by a complex pattern of multiple market actors, interconnections and dynamics in the entire supply chain, which are widely non-transparent and still intangible to allow for reliable, long-term project and investment planning. This study is addressing this research gap by reviewing the current state of worldwide pellet trade economics and price risks based on three case studies on pellet trade from Canada, Australia and Russia to the EU.

Keywords EU energy targets • Biomass supply • Reliability • Investment planning

Biomass plays a major role to fulfil the EU's energy targets for 2020 (EP, EU Council 2009). The EU's 20-20-20 targets aim for a 20 % reduction of greenhouse gas emissions from energy, a 20 % increase in efficiency and a 20 % increase of renewable energy sources in energy consumption by 2020 compared to 1990 levels. So far, renewables take a share of around 13 % of the EU27's energy consumption, whereas most comes from woody biomass. Not only today, but also for the 2020 targets biomass should contribute much more than 50 % to the EU's renewable energy consumption, and 19 % (or 16 % solid biomass) to the EU's renewable electricity production (Beurskens et al. 2011; Donnelly 2012), see Fig. 1.1.

In 2011, around 3300 PJ primary energy of biomass was produced in the EU, whereas slightly more solid biomass (3383 PJ) was consumed. That means 66 % of the EU's primary renewable energy production comes already from biomass (Eurostat 2013a). Thereby, currently 72 % of biomass is used for heating and cooling, about 15 % for transportation and 13 % for electricity use. In 2010, most electricity from biomass was produced in Germany with 30,000 GWh/a, followed by Sweden, the UK, and Finland. For this electricity, forestry in terms of wood

R. Ehrig et al., *Economics and Price Risks in International Pellet Supply Chains*, SpringerBriefs in Applied Sciences and Technology, DOI: 10.1007/978-3-319-07016-2_1,

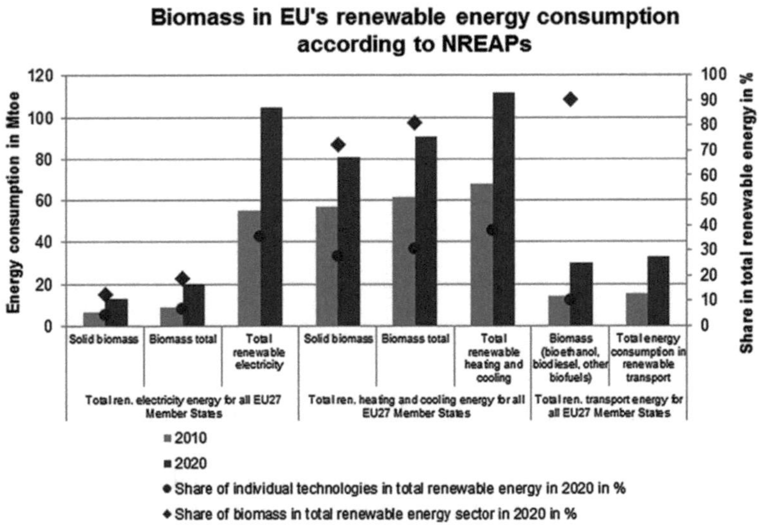

Fig. 1.1 Contribution of biomass to EU27 renewable energy consumption in 2010 and 2020 according to National Renewable Energy Action Plans (NREAPs). *Source* Beurskens et al. (2011)

and wood waste is the main supply sector. According to EU estimations, forestry will remain the main supply sector for the EU's solid biomass supply by 2020 (Donnelly 2012).

For reaching the ambitious energy targets, the EU member states will rely on biomass imports from non-EU countries, in particular for electricity generation (Hewitt 2011). Co-firing of industrial wood pellets represents one of the most cost-efficient and easy-to-adopt technologies to produce renewable electricity. It is widely implemented throughout EU countries like the Netherlands, Belgium, UK and Scandinavia. According to Lamers et al. (2012), EU imports of industrial pellets already reached 2.5 million ton in 2010, which amounts to more than 20 % of the EU consumption. Most imports are from Canada with almost 1 million ton pellets in 2010, then 0.7 million ton from the USA, 0.4 million ton from Russia and 63,000 ton from Australia. Junginger (2012) estimates that the EU pellet demand will rise to between 20 and 50 million ton by 2020, which means a sharp increase of almost 16 million biomass imports to the EU.

In this frame, the present work reviews the current state of worldwide pellet trade economics based on three case studies on pellet trade from Canada, Australia and Russia to the EU. With these, the following questions are raised:

(1) Which price fluctuations have the biggest influence on pellet supply costs and thus on the import price? How do they affect electricity production costs during co-firing?
(2) What is the magnitude of pellet price volatility over typical supply contract periods?
(3) In which way do pellet importers face and handle price variations and de-risk the supply chain within contractual relationships?

The present work is organised as follows: Methods and related works are presented in Chap. 2. Typical supply costs along three real case studies are assessed in Chap. 3. Most crucial price variations along the pellet supply chains are evaluated in Chap. 4. Resulting findings are discussed in Chap. 5. Finally, answers and conclusions on the central questions are given in Chap. 6. Detailed data on technology parameters and prices can be found in the Appendices.

References

Beurskens LWM, Hekkenberg M, Vethman P (2011) Renewable energy projections as published in the national renewable energy action plans of the European member states, summary report. Energy Research Centre of the Netherlands and European Environment Agency, Copenhagen

Donnelly M (2012) Biomass—role in achieving the 20 % target. Paper presented at the European biomass power generation conference, London, 1–2 Oct 2012

European Parliament and the Council of the European Union (2009) Directive 2009/28/EC of the European Parliament and of the Council of 23 April 2009 on the promotion of the use of energy from renewable sources and amending and subsequently repealing Directives 2001/77/EC and 2003/30/EC, L140/ 16–62, Brussels, Strassburg

Ehrig R, Wörgetter M, Pointner C, Kristöfel C, Strasser C (2011) Biomass mobilisation for industrial-scale bioenergy plants. Practical approach for establishing real biomass supply pathways in Austria. In: Proceedings of the 19th European Biomass Conference and Exhibition in Berlin, 6–10 June 2011 (Eta Florence/ WIP Munich, Florence, Munich, 2011)

European Commission, Eurostat (2013a) EU27 trade since 1995 by CN8, Monthly data, Brussels, last update 27.04.2012

Hewitt J (2011) Flows of biomass to and from the EU. An analysis of data and trends, report published by FERN, Brussels, 2011

Junginger M (2012) Overview of global solid and liquid biomass trade for energy. In: Proceedings of IEA Bioenergy Conference 2012, Vienna, Nov 13–15, 2012

Lamers P, Junginger M, Hamelinck C, Faaij A (2012) Developments in international solid biofuel trade—an analysis of volumes, policies, and market factors. Renew Sustain Energy Rev 16(5):3176–3199

Pöyry Ed (2011) Pellets—Becoming a Global Commodity? Pöyry View Point. Global market, players and trade to 2020, Vantaa

Chapter 2
Methods and Related Work

Abstract This chapter gives an overview of related research and outlines the methods applied in the present study. The study is analysing three real case biomass supply chains with an in-depth assessment of individual variables underlying actual market actions. Because of increasing biomass resources and recent dominance over pellet imports into the European market, the considered origin countries are Western Canada, Western Australia, and Northwest Russia. Studied supply phases include the raw material production and delivery, pellet production, transport to Europe, as well as delivery and conversion in a coal based co-firing power plant in the EU. The specific supply costs from origin country to the EU are derived from current market and country related data. For evaluating the pellets production and end-conversion in power plants, a full cost account is applied. For most vulnerable, market-related cost items the imputed risk is evaluated as effect of underlying price changes in a 3- to 10-year period. Corresponding de-risk strategies are concluded from expert interviews.

Keywords State of research • Case study approach • Cost account • Risk evaluation

2.1 State of Biomass Supply Chain Research

Since the last 10–15 years, several studies have been dealing with modelling and optimisation of particular biomass supply chains using GIS models, linear or mixed integer modelling (see e.g. Freppaz et al. 2004). These all have a specific focus (e.g. optimisation of logistics, costs or allocation of resources), and respond to a given framework and several assumptions. That is a local logistics network,

R. Ehrig et al., *Economics and Price Risks in International Pellet Supply Chains*,
SpringerBriefs in Applied Sciences and Technology, DOI: 10.1007/978-3-319-07016-2_2,

specific transportation means or the allocation of resources for specific end-use demand. Though, these models represent the actual market situation only little.

More focussed on actual trade flows, a comprehensive model on biomass supply chains was accomplished by Hamelinck et al. (2005) comparing different international bioenergy chains to Europe with focus on logistics. An evaluation of supply costs from Argentina to the Netherlands was done by Uasuf (2010). Costs for the pellet supply from British Columbia to the EU has been assessed before by Sikkema et al. (2010). In several market reports, the given framework for international biomass trade and specific aspects like shipping (Bradley et al. 2009) or equity and investments (Bradley 2010) are discussed. Moreover, in a scenario-based study Heinimö (2011) determined critical factors for the future development of the global (solid) biomass market (Heinimö 2011):

- Price competitiveness of bioenergy
- Energy policy (subsidies, R&D)
- Imbalance between supply and demand of bioenergy (sources)
- International agreements
- Sustainability issues to the utilisation of biomass

These existing studies serve as profound background and for comparison of assessed supply chains in this thesis. Nevertheless, so far there has been hardly any study, which combines an analysis of real case biomass supply with a detailed assessment of individual variables underlying actual market, regulatory and technology actions.

2.2 Case Study Compilation

Three different case studies for pellet imports to Europe are investigated for associated supply costs from resource origin to end-user in the EU, following the pattern in Fig. 2.1. Studied phases include the raw material production and delivery, pellet production, transport to Europe as well as delivery and conversion in a coal-based co-firing power plant, located 75 km from EU import harbour. Because of increasing biomass resources and recent dominance over pellet imports into the European market, British Columbia (Canada), Western Australia, and Northwest Russia are chosen as the case studies (Junginger 2012; Lamers et al. 2012; Röder 2012). They further offer a good comparison as they differ significantly in biomass source, distance and region.

Related work was accomplished by Hamelinck et al. (2005) comparing different international bioenergy chains to Europe with focus on logistics. An evaluation of supply costs from Argentina to the Netherlands was done by Uasuf (2010). Costs for the pellet supply from British Columbia to the EU has been assessed before by Sikkema et al. (2010), who already discussed price sensitivities. These existing studies serve as profound background and for comparison of assessed supply chains in this work. The present study reassesses the Canadian case

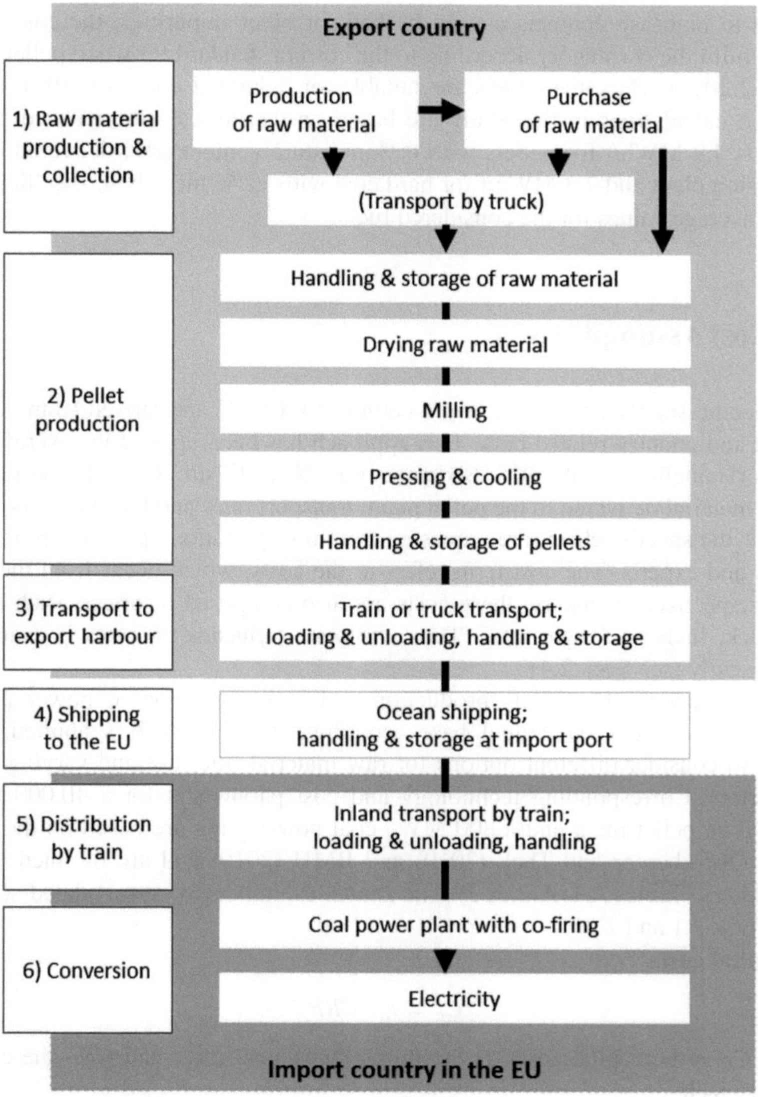

Fig. 2.1 Outline of pellet supply chain model

because of its prominent role in pellet exports to the EU, in order to vary input parameters and with the new target to explore price risks along the supply chain.

The pellet production phase and logistic operations are based on typical capacities and on technology in use in the respective countries. For all chains, two fuel options are distinguished for drying the raw material: biomass (Bio) and natural gas (NG). As result, a detailed description of the three pellet supply cases from resource origin to conversion plant in Europe is presented in Sects. 3.1–3.3.

Due to high ash contents caused by bark or other impurities, the considered pellets fulfil the B category according to the current standard for wood pellets (EN 14961-2:2011). Thus, the pellets are suitable for industrial use only. All information and calculations in this study are based on the net calorific value of fuels, which is 4.9 MWh/t for pellets with 6 % moisture content (mc) delivered at the conversion plant and 7.8 MWh/t for hard coal with <2 % mc. These specifications reflect average values for the considered fuels.

2.3 Cost Assumptions

The specific supply costs from origin country to the EU are derived from current market and country-related data. This approach has been applied in several other studies (Hamelinck et al. 2005; Sikkema et al. 2010; Uasuf 2010). Thereby, costs of raw material delivered to the pellet plant, transport rates and logistic costs, common for the specific chain, are requested from transportation operators, bioenergy traders and experts. The cost term refers to the costs, which occur from the end-user perspective. By nature, these costs are also composed of prices, such as for feedstock, fuels or freight rates. Thus, the price influence on costs is examined subsequently (see Sect. 2.4).

For evaluating the pellets production and end-conversion in power plants, a full cost account on annual base according to VDI 2067 is applied. This allows to consider different options for raw material, fuel use and varying other parameters. Corresponding technology and cost parameters for a 40,000 t/a or 120,000 t/a pellet plant and a 800 MW_{el} coal power plant are based on the studies by Obernberger and Thek (2010) and BMU (2010) and are outlined in the Appendices (Tables A.1, A.2, A.3). The annual capital costs are calculated according to Eqs. 2.1 and 2.2.

Capital costs

$$C_c = I_0 \cdot CRF \tag{2.1}$$

where C_C is the capital costs, I_0 the Initial investment costs and *CRF* the capital recovery factor.

Capital recovery factor

$$CRF = \frac{(1+i)^n \cdot i}{(1+i)^n - 1} \tag{2.2}$$

where i is the interest rate of the project, and n is the lifetime of equipment.

The electricity production costs (without heat extracts) are calculated according to Eq. 2.3.

Levelised costs of electricity

$$C_{el} = \frac{C_C + OM}{E} + \frac{C_F}{\eta \cdot 0.0036} + CO_2 \text{ costs} \tag{2.3}$$

Table 2.1 Currency exchange rates to Euro for the year 2011

1 € corresponds to	Currency
1.35	Australian dollar: AUS-$
1.38	Canadian dollar: CAN-$
40.88	Russian rouble: RUB
1.39	US dollar: US-$

Source Eurostat (2012a)

where C_{el} are the levelised costs of electricity [€/kWh], *OM* are the annual costs for operation, maintenance and other costs [€/a], calculated as relative share (%) of investment costs, *E* is the annual electricity production [kWh], C_F are the annual fuel costs [in € per primary GJ], η is the efficiency of the plant [%], and CO_2 costs are the charged EU emission allowances for combusting the used fuel [€/kWh].

In this work a 10 % co-firing of pellets (80 MW_{el} installed biomass capacity) is regarded as technically viable. Assumed extra costs due to co-firing pellets can be found in the Appendices (Table A.2).

The underlying cost data for energy conversion are adjusted using current fuel prices (EEX 2012a, b). The allocated CO_2 emissions from combustion of hard coal are calculated according to the EU's emission trading system (EC 2007). All calculations are estimated in € using the exchange rates for 2011 (see Table 2.1). VAT, profit margins or supply charges are not included. The results are presented either in €/t pellets (delivered at import harbour or conversion plant) or in €/MWh_{el} converted energy. The resulting structuring of cost data along the supply chain in Sect. 3.1–3.3 is inspired by Sikkema et al. (2010).

2.4 Evaluation of Price Risks

2.4.1 Methodology for Evaluating 10-year Price Variations

The defined supply costs are basis for evaluating market risks in the supply chain. Thus, for most vulnerable, market-related cost items the imputed risk is evaluated as effect of underlying price changes. This is a common approach during cost accounting for entrepreneurial activities. That means, based on historical price variations within one contractual period, the expected losses or revenues for the future period can be extrapolated (Mumm 2008). Certain attempts of this approach have been performed for individual price effects by Sikkema et al. (2010) or in connection with sensitivity analyses by Uasuf (2010). Sikkema et al. (2011) further explored the general market and trade conditions and prospects of the European pellets market. Anyway, so far there has been no comprehensive price risk analysis for pellet supply chains.

First, most relevant price variations within the recent 10 years are identified and determined. The 10-year period covers the time frame the pellet market has just

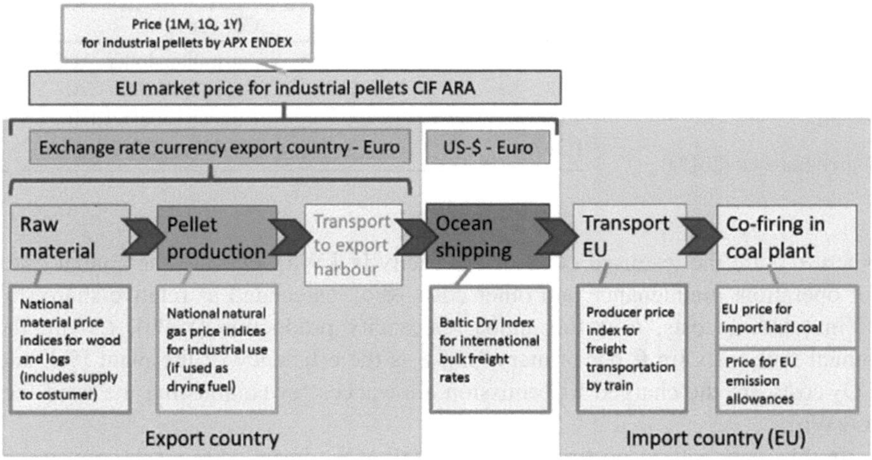

Fig. 2.2 Considered price indices for modelling price fluctuations along pellet supply chains destined for EU co-firing

evolved. Hence, a straightforward and unambiguous statistical analysis is applied by assessing the range of price variations as factors of total supply chain costs.

As shown in Fig. 2.2, different price indices along the supply chain have been identified. The stated data series for a 10-year period are inflation-adjusted using the relevant consumer price indices (Eurostat 2012b).

Based on the evaluated costs for Canadian, Australian and Russian pellets exported to the EU, the total supply costs free conversion plant are assumed to be arithmetic mean. With that, the relevant price variation—in terms of their standard variation or range—is charged as multiplier of the respective cost share in the supply chain (see Eqs. 2.4 and 2.5). In that way, the price effect of each factor on total costs can be revealed.

Supply costs subject to standard variation of price index xp

$$\boldsymbol{C}_{\boldsymbol{Total}(x,\sigma_{xp}^{+/-})} = \boldsymbol{C}_x \cdot \left(\boldsymbol{1} + \boldsymbol{\sigma}_{xp}^{+/-}\right) + \sum_{i=1,i\neq x}^{n} \boldsymbol{C}_i \tag{2.4}$$

where $C_{Total(x,\sigma_{xp}^{+/-})}$ are the total supply costs, which are subject to the standard deviation of price index xp in cost item C_x[€/t], $\sigma_{xp}^{+/-}$ is the negative (−) or positive (+) standard deviation of index xp, described as percentage of arithmetic mean of index xp [%] and C_i are the cost components 1 to n in the supply chain [€/t].

Supply costs subject to lower and upper range limit of price index xp

$$\boldsymbol{C}_{\boldsymbol{Total}(x,R_{xp}^{+/-})} = \boldsymbol{C}_x \cdot \boldsymbol{R}_{xp}^{+/-} + \sum_{i=1,i\neq x}^{n} \boldsymbol{C}_i \tag{2.5}$$

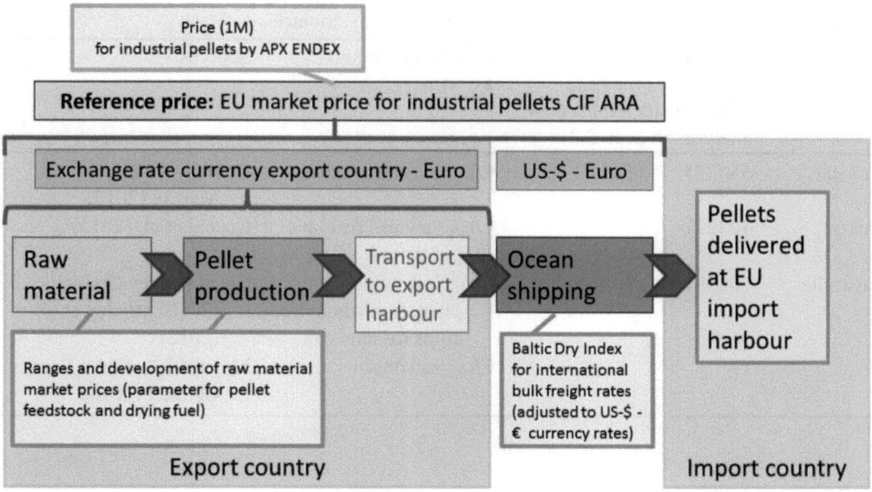

Fig. 2.3 Considered indices and prices for simulating 3-year-price fluctuations

where $\boldsymbol{C}_{\boldsymbol{Total}(x,R_{xp}^{+/-})}$ are the total supply costs, which are subject to the lower $\left(R_{xp}^{-}\right)$ or higher $(R_{xp}{}^{+})$ range of price index xp in cost component x.

The corresponding results are described in Sect. 4.2.

2.4.2 Methodology for Evaluating Recent 3-year Price Variations

After the observation of long-term price variations, actual price changes in the 3-year period from 2008 to 2011 are investigated. A period of 3 years complies with the typical (long-term) planning and contracting horizon of pellet producers and end-users (Alakangas et al. 2012; Sikkema et al. 2011).

The simulation of the recent 3-year variations reflects the cumulative annual price changes due to raw material prices, exchange rates and ocean shipping rates as the most crucial price factors (see Fig. 2.3). Apart from that, all other costs are assumed to be constant as defined for 2011, the "base price" (see Sect. 3.4). The EU pellet market price for imported pellets reported by APX (2012) serves as reference.

For actual pellet feedstock price variations (see Sect. 4.1.1. for explanation) the use of alternative assortments is considered. Due to higher quality, the delivery over long distances or demand from other industries, this feedstock is associated with higher purchase prices, see Table 2.2. For Australia, the case of cheaper alternative sawdust at minor costs is investigated as well (May 2012).

The results of the 3-year price simulation are displayed in Sect. 4.3.

Table 2.2 Price variation of raw material in the case study countries

Country	Price changes (in brackets the base price considered for cost analyses in Sect. 2.3)	Unit and biomass assortment	Sources
Canada	65 (32)	€/t_{dry} harvest residues	Bradley (2012), Murray (2012)
Russia	40 (22)	€/t_{dry} sawmill residues purchased	Cocchi et al. (2011)
Australia	70 (39)	€/t_{dry} low value plantation whole tree chips (assumed)	Clean Energy Council (2010) Stucley et al. (2012)
	11	€/t_{dry} sawdust	May (2012), Clean Energy Council (2010)

2.4.3 Expert Interviews on Supply Risks and De-risk Strategies

The risk analysis is complemented by personal communications with pellet market actors and related literature regarding their evaluation and hedging mechanisms against price risks in international biomass trade. The interviews were conducted face-to-face or via e-mail, following a semi-structured guideline. For confidentiality reasons, interview results containing potential sensitive data are indicated in aggregated form (Interviews 2011–2013). With these, insight is gained into possible hedging and contractual provisions to catch arising price gaps. Resulting findings can be found in Sects 4.1 and 4.4.

References

Alakangas E, Junginger M, van Dam J, Hinge J, Keränen J, Olsson O, Porsö C, Martikainen A, Rathbauer J, Sulzbacher L, Vesterinen P, Vinterbäck J (2012) EUBIONET III – Solutions to biomass trade and market barriers. Renew Sustain Energ Rev 16(6):4277–4290

APX-ENDEX (2012) Historical pellets market prices, Weekly prices. Compilation by Sipke Veer, Amsterdam, 14.07.2012

Bradley D (2010) Canada Report on Bioenergy 2010, Sponsored by Canadian Bioenergy Association, Natural Resources Canada, Canadian Wood Fibre Centre, Wood Pellet Association of Canada, Ottawa, 15 Sept 2010

Bradley D (2012) Canada—biomass supply/demand, export availability. In: Proceedings of world bioenergy 2012, Jönkoping, Sweden, 30 May 2012

Bradley D, Diesenreiter F, Wild M, Tromborg E (2009)World Biofuel Maritime Shipping Study. Report accomplished for IEA Task 40. Ottawa, Vienna, 01 July 2009

Clean Energy Council (2010) Bioenergy industry, report prepared by stephen schuck. Killara, Australia, June 2010

Cocchi M, Nikolaisen L, Junginger M, Sheng Goh S, Heinimö J, Bradley D, Hess R, Jacobson J, Ovard LP, Thrän D, Hennig C, Deutmeyer M, Schouwenberg PP, Marchal D (2011) Global wood pellet industry market and trade study, prepared for IEA Bioenergy Task 40: International sustainable bioenergy trade, Florence, Dec 2011

European Commission (2007) Commission decision of 18 July 2007 establishing guidelines for the monitoring and reporting of greenhouse gas emissions pursuant to Directive 2003/87/EC of the European Parliament and of the Council. (2007/589/EC), Official Journal of the European Union, L 229/1, Brussels, 31 Aug 2007

European Commission, Eurostat (2012a) Euro/Ecu Exchange rates 2000—2011. Quarterly data, Brussels, last update 04 July 2012

European Commission, Eurostat (2012b) Harmonised Consumer Price Index—Inflation rate 1997—2011 European and Non-European countries, Brussels, 2012

European Energy Exchange (2012a) ARA Coal Year Futures, Request of historical prices from 2006 to Oct 2012 at: http://www.eex.com/de/Marktdaten/Handelsdaten/Kohle/Coal. Accessed 15 Oct 2012

European Energy Exchange (2012b) Prices and trade volume of EU Emission Allowances, Request of historical prices and volumes at: http://www.eex.com/de/Marktdaten/Handelsdaten/Emissionsrechte. Accessed 15 Fib 2012

Freppaz D, Minciardi R, Robba M, Rovatti M, Sacile R, Taramasso A (2004) Optimizing forest biomass exploitation for energy supply at regional level. Biomass Bioenergy 26:15–25

Hamelinck C, Suurs R, Faaij A (2005) International bioenergy transport costs and energy balance. Biomass Bioenergy 29(2):114–134

Heinimö J (2011) Developing markets of energy biomass—local and global perspectives. Ph.D. thesis at Mikkeli University, Mikkeli, Finland

Interviewed pellet producers, traders and end-users (2011–2013) Personal communication with: Dusan S (2011), Black pellets manager of Vattenfall Europe AG, 08 Aug 2011. Hermes HD (2012) Director Business Development Biomass of Vattenfall Europe GmbH, 29 Oct 2012. Mertens J (2013) Biomass Procurement Officer at GDF Suez, 07 Feb 2013. Lugner M (2012) Sales Manager Max Lugner from pellets producer Schweighofer, 15 Jun 2012. Pease H (2013) Senior Biofuel Portfolio Manager for RWE Supply and Trading, 18 Aprl 2013

Junginger M (2012) Overview of global solid and liquid biomass trade for energy. In: Proceedings of IEA Bioenergy Conference 2012, Vienna, 13–15 Nov 2012

Lamers P, Junginger M, Hamelinck C, Faaij A (2012) Developments in international solid biofuel trade—An analysis of volumes, policies, and market factors. Renew Sustain Energ Rev 16(5):3176–3199

May B (2012) Personal communication about biomass prices and the australian biomass market with former researcher at commonwealth scientific and industrial research organisation (CSIRO). Aust Natl Sci Agency 06(08):2012

Mumm M (2008) Kosten- und Leistungsrechnung: Internes Rechnungswesen für Industrie- und Handelsbetriebe. Springer, Hamburg

Murray G (2012) Personal e-mail communication with executive director of the Wood Pellet Association of Canada, 22 Oct 2012

Obernberger I, Thek G (2010) The pellet handbook. The production and thermal utilisation of biomass pellets, Earthscan, London

Röder H, (2012) Global development of bioenergy, presentation from Pöyry consulting at University of Applied Sciences Wiener Neustadt. Wieselburg, Nov 2012

Sikkema R, Junginger M, Pichler W, Hayes S, Faaij A (2010) The international logistics of wood pellets for heating and power production in Europe: Costs, energy-input and greenhouse gas balances of pellet consumption in Italy, Sweden and the Netherlands. Biofuels Bioprod Biorefinery 4(2):132–153

Sikkema R, Steiner M, Junginger M, Hiegl W, Hansen MT, Faaij A (2011) The European wood pellet markets: current status and prospects for 2020. Biofuels Bioprod Biorefinery 5(3):250–278

Stucley C, Schuck S, Sims R, Bland J, Marino B, Borowitzka M, Abadi A, Bartle J, Giles R, Thomas Q (2012) Bioenergy in Australia, Status and Opportunities. Report, Surrey Hills, Victoria, Nov 2012

Uasuf A (2010) Economic and environmental assessment of an international wood pellets supply chain: a case study of wood pellets export from northeast Argentina to Europe. Doctoral thesis, Albert-Ludwigs-Universität Freiburg im Breisgau, Faculty of Forest and Environmental Sciences, Dec 2010

Umweltbundesministerium (ed) (2010) Leitstudie 2010—Langfristszenarien und Strategien für den Ausbau der erneuerbaren Energien in Deutschland bei Berücksichtigung der Entwicklung in Europa und global, Appendix II, Stuttgart, Kassel, Teltow, Dec 2010

Chapter 3
Pellet Supply Costs Along Three Case Studies

Abstract This chapter describes the three chosen case studies for biomass supply from Canada, Australia, and Russia to the EU. The case studies contain a detailed cost outline from biomass resource to final consumer for the year 2011. The supply patterns include raw material, pellet production phase, transportation and shipping to the EU, delivery to the conversion plant and final conversion in a coal co-firing plant. The individual costs are summarised and compared, explaining the related market connections. Dedicated and region specific cost drivers and economic framework conditions are defined along the whole biomass supply chain.

Keywords Case study analysis–Canada • Russia and Australia–supply costs

3.1 Canadian Pellets to Europe

British Columbia in Western Canada has a vast potential of 417 million ha forests representing the 3rd largest forest area in the world (Ferguson 2010). In 2010, the lumber production was 27 million m^3 (Bradley 2010) with resulting volume of sawmill residues between 18 and 25.6 million m^3/a (6.8–9.7 million t_{dry}/a) (Verkerk 2008; Wiik et al. 2009).

The raw material for pellet production taken into account are sawdust and shavings from spruce (36 % mc), which is transported 100 km on average from sawmills or harvesting sites to the pellet plant (Sikkema et al. 2010; Urbanowski 2005). The raw material costs are set 23.41 €/t feedstock (wet) delivered at pellet mill gate. This corresponds to 33.25 €/t_{wet} pellets, which is in line with the data provided by Sikkema et al. (2010) and Bradley (2012).

The modelled input data and results for the pellet production and delivery phases are listed in Table 3.1. The assumed pellet plant capacity is 120,000 t/a. Within the production process a rotary drum dryer with respective energy demand is considered (Magelli et al. 2009; Urbanowski 2005).

R. Ehrig et al., *Economics and Price Risks in International Pellet Supply Chains*,
SpringerBriefs in Applied Sciences and Technology, DOI: 10.1007/978-3-319-07016-2_3,

Table 3.1 Cost breakdown along the supply chain Canada–Europe

	Basic data	€/t pellets delivered	References
1) Raw material supply			
Raw material	Sawdust and shavings from spruce, 36 % mc on average	5.00 €/t feedstock	Bradley (2010), Urbanowski (2005)
Transport of raw material to pellet plant incl. Handling and storage	Average distance 100 km truck transport from saw mill incl. unload	18.41 €/t feedstock	Urbanowski (2005) with annual 2 % cost increase
Raw material delivered to pellet plant	170,438 t/a raw material required	33.25	Calculation in accordance to Ferguson (2010), Bradley (2012)
2) Pellet production			
Pellet production	120,000 t pellet production 1 % mass losses included	9.18 million € investment costs	Own calculation based on plant data from Obernberger and Thek (2010) (assuming the use of a rotary drum dryer based on Urbanowski (2005) and Magelli et al. (2009)) and local energy prices; incl. handling and storage of raw material and pellets
Capital costs	6 % interest rate, 17.5 average life time	7.18	
Consumption costs excl. raw material and fuel		7.52	
Natural gas costs for drying	90 % boiler efficiency	7.42	
Biomass costs for drying	90 % boiler efficiency	3.17	
Operation and maintenance	3 shifts/d; 7 working days/week	4.82	
Other costs		2.60	
Subsum pellet production costs	Dried with natural gas	62.79	
	Dried with biomass	58.54	

(continued)

Table 3.1 continued

	Basic data	€/t pellets delivered	References
3) Transport to export harbour			
Loading		1.50	Sikkema et al. (2010)
Train transport to export port	500 km, 84 t (130 m³) load per railcar	24.85	CN (2012), Spelter and McKeever (2009)
4) Ocean shipping to the EU			
Handling and storage	incl. 1 % mass losses	2.38	Sikkema et al. (2010)
Ocean transport	Handysize, 16.500 km; Vancouver-Rotterdam incl. 1.5 % mass losses	36.61	Ferguson (2010), Melin (2012)
Import costs Rotterdam (subtotal)	Pellets dried with natural gas	123.87	
	Pellets dried with biomass	128.12	
5) Distribution to conversion plant			
Handling and storage at import port	Discharge by clam buckets, conveyor system, storage, load	5.00	Sikkema et al. (2010), Sumetzberger (2012), Obernberger and Thek (2010)
Train transport to conversion plant BE or NL	75 km	5.50	Calculation based on Ökoinstitut and Enerko (2008), Prognos (2006)
Total costs free conversion plant	Pellets (NG)	138.62	
	Pellets (Bio)	134.37	

As seen in Table 3.1, the total costs for Canadian pellets delivered to Rotterdam are therefore 124–128 €/t, or 134–139 €/t pellets delivered to the conversion plant. The composition of costs is similar to those estimated by Ferguson (2010) and Sikkema et al. (2010). Compared to Sikkema et al. (2010) less freight costs have been assumed here, because increased freight capacity and return trips on the Vancouver—Rotterdam route could be realised within the last years (Murray 2012).

3.2 Australian Pellets to Europe

Southern Australia offers an increasing potential of eucalyptus (blue gum) plantations from marginal farm land destined for industrial pellets production. It is expected to provide significant volumes to the global pellet market, including Europe (Junginger 2012; Röder 2012). Foresters expect an extension of the plantation area from 0.58 million ha in 2009 up to 2 million ha in 2014. In 2010, the first large-scale 125,000 t/a pellet plant (later upscaled to 250,000 t/a) started the production of industrial wood pellets from plantation residues in Albany (Western Australia), mainly for export to Northwest Europe. The industry announced an increasing set up of pellet plant facilities to 850,000 t/a production capacity in near future (Smith 2010; Waring 2010). But different factors currently hinder the pellet export: The operator of the largest Australian pellet plant recently faced economic problems due to strength of Australian dollar to Euro and because of switching from residues to more expensive raw material (Grieve 2012). Another influencing factor for the Australian-European trade is the competition with Asia, which could lead to more exports from Australia to Japan or Korea (Waring 2010). Nevertheless, the EU is still one of the dedicated target markets for future pellet exports from Australia (Clean Energy Council 2010; Lamers et al. 2012; Lang 2013).

The present supply model is based on the evaluation of the existing pellet plant in Western Australia replying on local eucalyptus plantation residues. The production phase is based on 120,000 t/a capacity, which corresponds to approximately one unit in Albany. All assumptions and results are listed in Table 3.2. The resulting total supply costs are in line with the estimations by Smith (2010), taking into account the 2009 framework assumed in his study.

3.3 Pellets from Northwest Russia to Europe

Russia has vast wood reserves and a strong wood industry. The region Northwest Russia is favoured by direct access to the Baltic Sea. Sawmills in the Leningrad region surrounding St. Petersburg process more than 1 million m^3 wood per year (Karjalainen and Gerasimov 2010). The annual forest waste composes at least 100 million m^3, reported by Cocchi et al. (2011). The number of pellet production plants is constantly growing. The 800,000 t/a production capacity reported in 2008 (Rakitova et al. 2009) and the recent increase by the 1 million t/a pellet plant

Table 3.2 Cost breakdown along the supply chain Australia–Europe

	Basic data	€/t pellets delivered	References
1) Raw material supply			
Raw material	Sawdust and wood chips from blue gum plantation incl. delivery, 45 % mc on average	27.20 €/t feedstock	ABARES (2011), Clean Energy Council (2010), Stucley et al. (2012)
Raw material delivered to pellet plant	198,327 t/a feedstock used	44.95	Own calculation based on feedstock characteristics and conversion factors
2) Pellet production			
Pellet production	120,000 t pellet production; additional coarse grinding unit suitable for wood chips; incl. 1 % mass losses	11.29 million € investment costs	Own calculation based on plant data from Obernberger and Thek (2010) and local energy prices; incl. Handling and storage of raw material and pellets
Capital costs	6 % interest rate, 17.5 average life time	8.83	
Consumption costs	excl. raw material and fuel	9.19	
Natural gas costs for drying	90 % boiler efficiency	13.83	
Biomass costs for drying	90 % boiler efficiency	8.27	
Operation and maintenance	3 shifts/d; 7 working days/week	5.25	
Other costs		2.60	

(continued)

Table 3.2 (continued)

	Basic data	€/t pellets delivered	References
Subsum pellet production costs	Dried with natural gas	84.65	
	Dried with biomass	79.09	
3) Transport to export harbour			
Loading		3.00	Obernberger and Thek (2010)
Truck transport to export port	40 km (20 km with empty return trip)	6.17	Allen (2011), Freight Metrics (2012), Smith (2010), unloading from Obernberger and Thek (2010)
4) Ocean shipping to the EU			
Handling and storage	Dedicated wood pellets terminal available; incl. 1 % mass losses	2.70	Albany port 2012 (without investment costs)
Ocean transport	Handysize or Panamax; Albany-Rotterdam 21,570 km; incl. 1,5 % mass losses	47.50	Alderton (2012)
Import costs Rotterdam (subtotal)	Pellets dried with natural gas	144.03	
	Pellets dried with biomass	138.46	
5) Distribution to conversion plant			
Handling and storage at import port	Discharge by clam buckets, conveyor system, storage, load	5.00	Sumetzberger (2012), Sikkema et al. (2010), Obernberger and Thek (2010)
Train transport to conversion plant	75 km	5.50	Calculation based on Ökoinstitut and Enerko (2008), Prognos (2006)
Total costs free conversion plant	Pellets (NG)	154.53	
	Pellets (Bio)	148.96	

Table 3.3 Cost breakdown along the supply chain Russia–Europe

	Basic data	€/t pellets delivered	References
1) Raw material supply			
Raw material	sawdust incl. delivery, 55 % mc on average	13.89 €/t feedstock	Cocchi et al. (2011)
Raw material delivered to pellet plant	80,800 t/a raw material required	28.06	own calculation based on feedstock characteristics and conversion
2) Pellet production			
Pellet production	40,000 t pellet production	3.74 million € investment costs	Own calculation based on plant data from Obernberger and Thek (2010) and local energy prices; incl. Handling and storage of raw material and pellets
Capital costs	6 % interest rate, 17.5 average life time	8.78	
Consumption costs excl. raw material and fuel		8.44	
Natural gas costs for drying	90 % boiler efficiency	8.80	Natural gas and electricity prices from Federal Tariff Service (2013), Gerasimov (2012)
Biomass costs for drying	90 % boiler efficiency	8.33	
Operation and maintenance	3 shifts/d; 7 working days/week	6.22	
Other costs		2.60	
Subsum pellet costs	Dried with natural gas	62.89	
	Dried with biomass	62.43	
3) Transport to export harbour			
Loading		1.50	Sikkema et al. (2010)
Train transport to export port	400 km	10.03	Karjalainen and Gerasimov (2010); in accordance with Rakitova and Kholodkov (2009)

(continued)

Table 3.3 (continued)

	Basic data	€/t pellets delivered	References
4) Ocean shipping to the EU			
Handling and storage	incl. 1 % mass losses	12.00	Cocchi et al. (2011), Ovsianko (2006)
Ocean transport	assumed 4,000 t capacity; St. Petersburg-Rotterdam; incl. 1.5 % losses	20.30	Sumetzberger (2012)
Import costs Rotterdam (subtotal)	Pellets dried with natural gas	106.72	
	Pellets dried with biomass	106.26	
5) Distribution to conversion plant			
Handling and storage at import port	Discharge by clam buckets, conveyor system, storage, load	5.00	Sumetzberger (2012), Obernberger and Thek (2010), Sikkema et al. (2010)
Train transport to conversion plant	75 km	5.50	Calculation based on Ökoinstitut and Enerko (2008), Prognos (2006)
Total costs free conversion plant	Pellets (NG)	117.22	
	Pellets (Bio)	116.76	

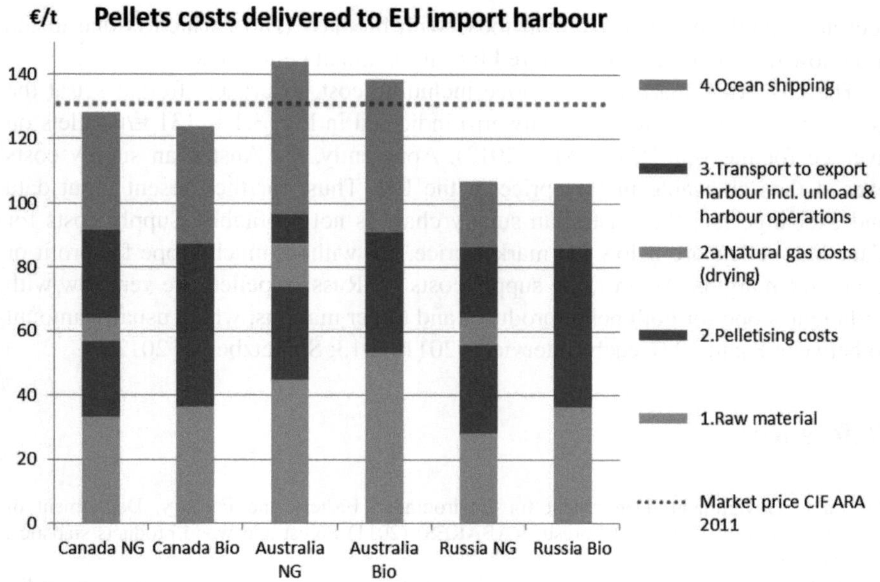

Fig. 3.1 Supply costs of defined pellet chains from origin to EU import harbour Rotterdam and average market price for pellets 2011. *Source* for pellets market price: APX (2012)

Vyborgskaya amount to a total capacity of 1.8 million t/a in the region. Up to now, the actual production is estimated at only 1 million t/a (Rakitova 2011). Most pellets are exported to the EU industrial market, where the revenue is comparably high. It is assumed that no export duties are imposed to exported pellets (Rakitova et al. 2009).

For the Russian supply case, sawdust from soft wood with 55 % mc is considered. The supplying wood industry is nearby the pellet plant. A smaller pellet plant with 40,000 t/a capacity is taken into account, which is the case at the installed sites in Northwest Russia (Cocchi et al. 2011). The distance to export harbour St. Petersburg is assumed to be 400 km. The costs for handling and transshipment of pellets at St. Petersburg harbour is high with 12 €/t pellets. This is due to a lack of a specialised cargo handling terminal with dedicated equipment for pellet transshipment (Cocchi et al. 2011; Ovsianko 2006; Rakitova and Kholodkov 2009). All input data and resulting costs are listed in Table 3.3.

3.4 Summary of Pellet Supply Costs

Based on the previous assessments (Sect. 3.1–3.3), the overall pellet supply costs for supply chains from Canada, Australia and Russia to EU import harbour are summarised in Fig. 3.1. The raw material costs amount to about one third of the total costs. Transport costs from pellet plant to EU import harbour amounts to another 30–50 % of total supply costs. The cases with natural gas (NG) as drying fuel are slightly more costly than those with biomass (Bio), which is due to the quite low raw material costs required for international pellet trade.

The CIF ARA market price (price including cost, insurance, freight set at the ports Amsterdam, Rotterdam, Antwerp) indicated in Fig. 3.1 is 131 €/t pellets on average for the year 2011 (APX 2012). Apparently, the Australian supply costs exceed the achievable market price in the EU. Thus, for the present input data and exchange rate the Australian supply chain is not profitable. Supply costs for Canadian pellets are below the market price, but without much scope for profit or guarantee margins. Against, the supply costs for Russian pellets are very low with sufficient scope for both pellet producer and trader margins, which usually amount to between 8 and 10 % each (Interviews 2011–2013; Sumetzberger 2012).

References

Australian Government, Department for Environment, Fishery and Forestry, Department of Agriculture, Fisheries and Forestry (ABARES) (2011) Forest and wood products statistics. Canberra. Dec 2011

Albany Port Authority (2010) Port Talk. Albany Port Community Newsletter. Albany, Australia. March 2010

Alderton P (2012) Personal communication with bulk parcel operator Furness withy Australia about ocean transport freight costs for wood pellets. 23 July 2012

Allen D (2011) Energy pellet production experience, presentation of plantation energy at CRC for forestry workshop: generating renewable energy from forest residues: opportunities, challenges and the latest research, Canberra, Australia, 14 Apr 2011

APX-ENDEX (2012) Historical pellets market prices. Weekly prices, Compilation by Sipke Veer, Amsterdam, 14 July 2012

Bradley D (2010) Canada Report on Bioenergy 2010, Sponsored by canadian bioenergy association, natural resources Canada, Canadian Wood Fibre Centre, Wood Pellet Association of Canada, Ottawa, 15 Sept 2010

Bradley D (2012) Canada—biomass supply/demand, export availability. In: Proceedings of world bioenergy 2012, Jönkoping, Sweden, 30 May 2012

Clean Energy Council (2010) Bioenergy industry, report prepared by stephen schuck, Killara, Australia, June 2010

Canadian National Railway Company (CN) (2012) Price proposal for appr. 500 km rail transport of pellets to port of Vancouver by covered hopper. http://www.cn.ca. Accessed 13 Dec 2011

Cocchi M, Nikolaisen L, Junginger M, Sheng Goh S, Heinimö J, Bradley D, Hess R, Jacobson J, Ovard LP, Thrän D, Hennig C, Deutmeyer M, Schouwenberg PP, Marchal D (2011) Global wood pellet industry market and trade study, prepared for IEA Bioenergy Task 40: International sustainable bioenergy trade, Florence, Dec. 2011

Federal Tariff Service Russia (2012) Natural gas wholesale price for industrial consumers for 2010. http://www.fstrf.ru/eng/tariffs/analit_info Accessed 20 Jun 2012

Ferguson S (2010) Bioenergy development in British Columbia, Canada. Mutual exchange with Europe, presentation of the BC bioenergy network at 4biomass workshop, Vienna, 5th Oct 2010

Freight Metrics Pty Ltd (2012) Truck operating cost calculator. http://www.freightmetrics.com.au/CalculatorsRoad/TruckOperatingCost/tabid/104/Default.aspx Accessed 10 Oct 2012

Gerasimov Y (2012) Personal communication with Russian biomass expert at Finish research centre metla. 01 Feb 2012

Grieve O (2012) Albany biomass pellet plant winds up, ABC rural broadcast service, News release on 02.03.2012. http://www.abc.net.au/rural/news/content/201202/s3422461.htm Accessed 10 Oct 2012

Interviewed pellet producers, traders and end-users (2011–2013) Personal communication with: Dusan S (2011) Black pellets manager of Vattenfall Europe AG, 08 Aug 2011. Hermes HD (2012) Director business development biomass of vattenfall Europe GmbH, 29 Oct 2012. Mertens J (2013) Biomass procurement officer at GDF Suez, 07 Feb 2013. Lugner M (2012) Sales manager max lugner from pellets producer Schweighofer, 15 Jun 2012. Pease H (2013) Senior biofuel portfolio manager for RWE supply and trading, 18 Aprl 2013

Junginger M (2012) Overview of global solid and liquid biomass trade for energy. In: Proceedings of IEA bioenergy conference 2012, Vienna, 13–15 Nov 2012

Karjalainen T and Gerasimov Y (2010) Energy wood resources availability and delivery cost in Northwest Russia. In: Proceedings of the third international symposium on energy from biomass and waste in venice, Italy, 8–11 Nov 2010

Lamers P, Junginger M, Hamelinck C, Faaij A (2012) Developments in international solid biofuel trade—an analysis of volumes, policies, and market factors. Renew Sustain Energ Rev 16(5):3176–3199

Lang A (2013) Australia and New Zealand: a future pellet exporter? In: Presentation by the world bioenergy association board member for Australasia–Oceania at the European Pellet Conference 2013, Wels, 28 Feb 2013

Magelli F, Boucher K, Bi HT, Melin S, Bonoli A (2009) An environmental impact assessment of exported wood pellets from Canada to Europe. Biomass Bioenergy 33(3):434–441

Melin S (2012) Design of logistics for regular and torrefied pellets. Presentation at BC-Korea Bioenergy Collaboration, Vancouver, 20 Mar 2012

Murray G (2012) Personal e-mail communication with executive director of the wood pellet association of Canada, 22 Oct 2012

Obernberger I, Thek G (2010) The pellet handbook. The production and thermal utilisation of biomass pellets, Earthscan, London, Washington

Ökoinstitut and Enerko (ed) (2008) Vergleich von heizkraftwerksvarianten für die Stadtwerke Kiel. Technische, wirtschaftliche und ökologische Bewertung, final report, Aldenhoven/Berlin

Ovsianko A (2006) Possibilities and threats for European business in the Russian bio-energy sector under its current scondition, Russian Forestry Review, St. Petersburg, Jan 2006

Prognos (2006) Variantenvergleich küste versus binnenland. ein volkswirtschaftlicher vergleich der kosten, versorgungssicherheit und umweltverträglichkeit von kraftwerksstandorten, Report, Berlin

Rakitova O, Kholodkov V (2009) The pellet market and wood resources in the North-West of Russia, Baltic 21 lighthouse project the baltic sea bioenergy promotion project, final report, St. Petersburg

Rakitova O, Ovsyanko A, Sikkema R, Junginger M (2009) Wood pellets production and trade in Russia, Belarus and Ukraine, market research report accomplished as subcontract for Pellets@las project, Utrecht, May 2009

Rakitova O (2011) The development of pellet production in Russia 2011, Presentation by National Bioenergy Union, The bioenergy international/Russia Infobio at EUBIONET workshop, Espoo, 19 Aprl 2011

Röder H (2012) Global development of bioenergy, presentation from Pöyry consulting at University of Applied Sciences Wiener Neustadt, Wieselburg, Nov 2012

Sikkema R, Junginger M, Pichler W, Hayes S, Faaij A (2010) The international logistics of wood pellets for heating and power production in Europe: Costs, energy-input and greenhouse gas balances of pellet consumption in Italy, Sweden and the Netherlands. Biofuels Bioprod Biorefinery 4(2):132–153

Sumetzberger H (2012) Personal communication with the supply and logistics manager of fuel trade company Genol GmbH and Co KG, Vienna, 08 Feb 2012

Smith D (2010) Australian pellet export outlook. In: Proceedings of IEA bioenergy exco65 workshop developing sustainable trade in bioenergy, held in conjunction with the meeting of the executive committee of IEA bioenergy in Nara City, Japan, 12 May 2010

Spelter H, McKeever T (2009) North America's wood pellet sector, research paper FPL–RP–656 by the United States department of agriculture, forest service, forest products laboratory, Madison

Stucley C, Schuck S, Sims R, Bland J, Marino B, Borowitzka M, Abadi A, Bartle J, Giles R, Thomas Q (2012) Bioenergy in Australia, status and opportunities, report, Surrey Hills, Victoria, Nov 2012

Urbanowski E (2005) Strategic analysis of a pellet fuel opportunity in Northwest British Columbia, Master thesis, University of Saskatchewan

Verkerk B (2008) Current and future trade opportunities for woody biomass end-products from British Columbia, Canada, master thesis at Utrecht University

Waring J (2010) Developing a pellet energy business in Australia, presentation at bioenergy Australia conference 2010, Sydney, 8–10 Dec 2010

Wiik C, Heiskanen VP, Kallio M, Antilla P (2009) Wood pellet raw material from British Columbia. Global potential of modern fuelwood, Research report VTT-R-10283-08, Jyväskylä

Chapter 4
Price Risks Along Pellet Supply Chains

Abstract This chapter deals with the evaluation of most crucial supply risks, which exist under real market conditions in the biomass market. General risk parameters and relevant indices along the biomass supply chain are discussed in detail. A modelling of 10-year-price variations based on the three investigated supply chains is followed and current 3-year-price volatilities are presented. Raw material prices, shipping rates, exchange rates, and the final market price turn out to be the main drivers for actual price variations and investment decisions. They can have a major impact on supply chain economics and investment success. Finally, responding hedging strategies for biomass supply chains and related investment decisions are summarised.

Keywords Biomass market • Supply risks • Price variations • Hedging strategies

The findings from Chap. 3 present the cost situation at a given time and in a specific configuration of supply chain options, and thus underlying various market parameters. Section 4.1 deals with the evaluation of most crucial supply risks, which exist under real market conditions. First, most crucial risk parameters and relevant indices along the supply chain are discussed in Sect. 4.1. Based on that, a modelling of 10-year-price variations based on the 3 investigated supply chains is followed in Sect. 4.2. Current 3-year-price volatilities are presented in Sect. 4.3. Responding hedging strategies are summarised in Sect. 4.4.

4.1 Price Risks and Indices Along the Pellet Supply Chain

4.1.1 Raw Material

Availability of feedstock is directly dependent on (regional) forests, on supplying sawmills as well as the regional demand for the specific feedstock. Hence, pellet

R. Ehrig et al., *Economics and Price Risks in International Pellet Supply Chains*, SpringerBriefs in Applied Sciences and Technology, DOI: 10.1007/978-3-319-07016-2_4,

producers compete not only among themselves, but also against the wood industry (panel and paper) for the same wood assortments (Ehrig et al. 2011; Interviews 2011–2013). Also, when not supplying sawmill residues to other industries, sawmills can convert it to heat and power for own utilisation. Bradley et al. (2010) argue that the economic recession led to worldwide closure of sawmills, which resulted in drastically reduced supply of low-cost sawdust, and in transition to more costly harvest residues as pellet feedstock. Actually, pellet feedstock is increasingly getting scarce while brokers enter this market to capture supply. One approach of pellet producers must therefore be to purchase feedstock from multiple sources and providers (Ehrig et al. 2011; Geiver 2012). Another approach is investing in energy crop plantations in warm climatic zones (Cocchi et al. 2011; Röder 2012).

Scarce data is available about biomass prices, because residues—the main feedstock for pellets—are not traded as commodity yet, and so market prices are not monitored. In the countries considered in this work, biomass residues from forests and industry are often available at no or minor costs, so harvesting and delivery are main cost factors (Interviews 2011–2013). Nevertheless, for a statistical indication national raw material indices for wood logs or the wood industry (ABARES 2011; FSSS 2012a; STATCAN 2012b, see Appendices B.1, B.3, B.5) are chosen as indicator for the price development for respective feedstock delivered to the consumer. That means any kinds of transportation efforts to the recipient are included here.

4.1.2 Pellet Production

For the pellet production process the main risks are the raise of new capital, the technology investment at a certain location and in time, as well as the cost-effective operation (Bradley et al. 2010; Interviews 2011–2013). These technical and entrepreneurial issues should not be considered here. As a measurable price risk natural gas as drying fuel in pellet production might have a considerable effect. Thus national price indices for natural gas are selected for investigation (FSSS 2012a; STATCAN 2012b; BREE 2012).

4.1.3 Transportation and Logistics

Transportation and logistics highly depend on the load volumes shipped, the frequency, utilised capacity of transportation routes and on the storage duration at transshipment sites, as discussed by Bradley et al. (2009), Senechal et al. (2009) and Sikkema et al. (2011). Mostly, the transportation over long distances is only economic when large volumes are shipped and return trips with the same means of transportation can be realised (Interviews 2011–2013; Rakitova et al. 2009). As

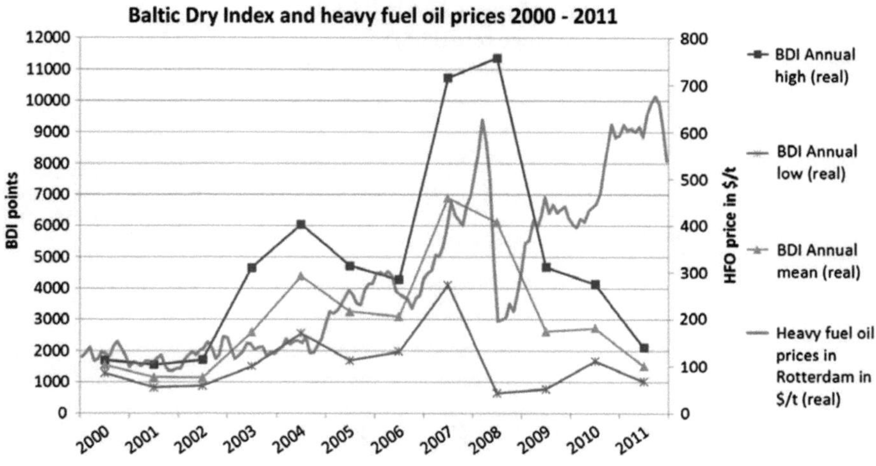

Fig. 4.1 Annual rates of Baltic Dry Index representing bulk shipping freight rates and heavy fuel oil prices from 2000 to 2011. *Sources* Maersk (2010), KMI (2012)

already mentioned in Chap. 3, the infrastructure for logistic operations is a crucial aspect with significant cost differences, depending on the availability of suitable port handling and loading equipment. Ocean shipping of pellets is operated by dry bulk transportation companies and generally underlies extremely fluctuating freight rates on a global scale. These variations influence operating costs for vessels, fuel, crews as well as the degree of matching supply and demand of transported goods, frequency of transport routes and available ship capacity (Bradley et al. 2009; Svanberg and Halldórsson 2013). For maritime shipping, the Baltic Dry Index (BDI) is the common price indicator. It records current ocean shipping rates (including heavy fuel oil) for transporting dry bulk good on main and most frequented trade routes, reported from various market actors (Maersk 2010; KMI 2012).

Figure 4.1 shows the inflation adjusted annual rates, peaks and lows of the BDI and heavy fuel oil prices from 2000 to 2011. Within this period, the index shows remarkable fluctuations with the total peak (11,793 points) and low (663 points), both in 2008. These are a result of the increasing global trade with goods, due to the enormous economic growth and trade in Asian countries and the lack of new shipping capacity to meet this demand. In particular, the strong fluctuation is also reflected by a variation of pellet shipping rates from Vancouver to Rotterdam between 35 and 100 US-$/t in 2009 (Bradley et al. 2009; Bradley 2010).

The actual freight rate for shipping pellets is usually fixed bilaterally in US dollar for every individual route, depending on volumes, utilised capacities and the contract period (Alderton 2012; Bradley 2010; Interviews 2011–2013; Sumetzberger 2012).

For the EU inland transport by train, no official indicator for price changes is available. Instead, the German transportation index for rail freight rates starting from 2006, serves as a reference for the further analysis (DESTATIS 2012, see Appendix, Figure B.8).

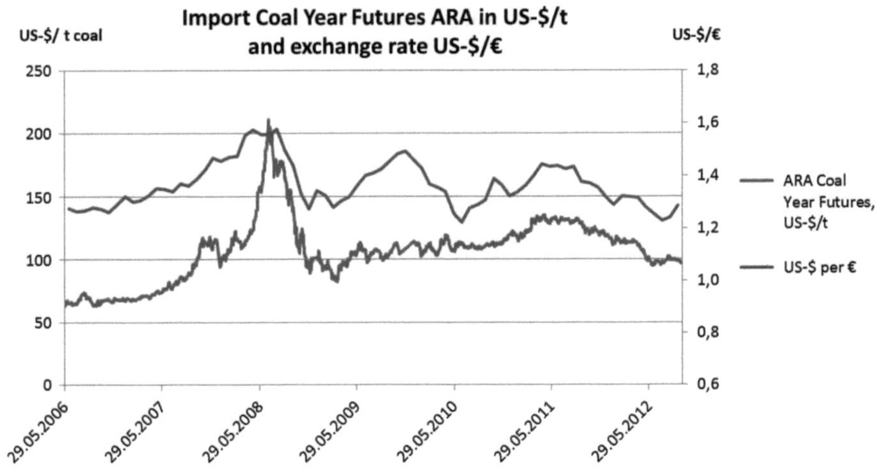

Fig. 4.2 Price of coal year futures (1 year forward prices) traded at ARA ports in US-$/t and exchange rate US-$/€. *Sources* EEX (2012b), Eurostat (2012a)

4.1.4 Conversion in Power Plants

Industrial pellets in Europe are currently destined for power production, above all for co-firing in coal power units. Dedicated funding opportunities are the crucial economic aspect for co-firing as investigated in Ehrig and Behrendt (2013). Other competing factors are the price for fossil fuel to be substituted by use of pellets and the charged EU emission allowances (Interviews 2011–2013).

The following analysis is based on the price developments for EU emission allowances and import hard coal (CIF ARA) reported by the energy exchanges EEX and bluenext from 2006 to 2012 (EEX 2012a; EEX 2012b; BLUENEXT 2012). As illustrated in Figs. 4.2 and 4.3, some correlation between coal prices and US-$/€ exchange rate as well as between prices of coal and EU emission allowances exists.

4.1.5 Price Factors Affecting the Whole Supply Chain

The upstream supply chain to export harbour is subject to varying exchange rates of the origin country's currency to Euro. The exchange rate data of AUS-$, CAN-$, US-$ and Russian rouble per € from the European Central Bank (Eurostat 2012a) are visualised in Figs. 4.4 and 4.5. For CAN-$ and AUS-$, a significant decline per € (or strong dollars to Euro) can be determined from 2009 to 2012. Based on average annual rates, the Australian dollar changed from 1.77 AUS-$/€ in 2009 to 1.35 AUS-$/€ in 2011, which amounts to 31 % variation. That means,

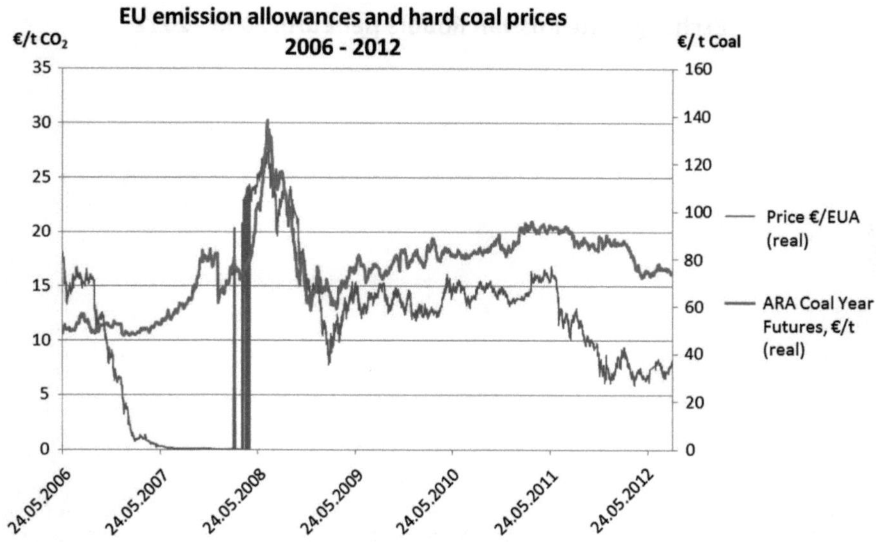

Fig. 4.3 EU emission allowances and German import price for hard coal. *Sources* BLUENEXT (2012), EEX (2012a)

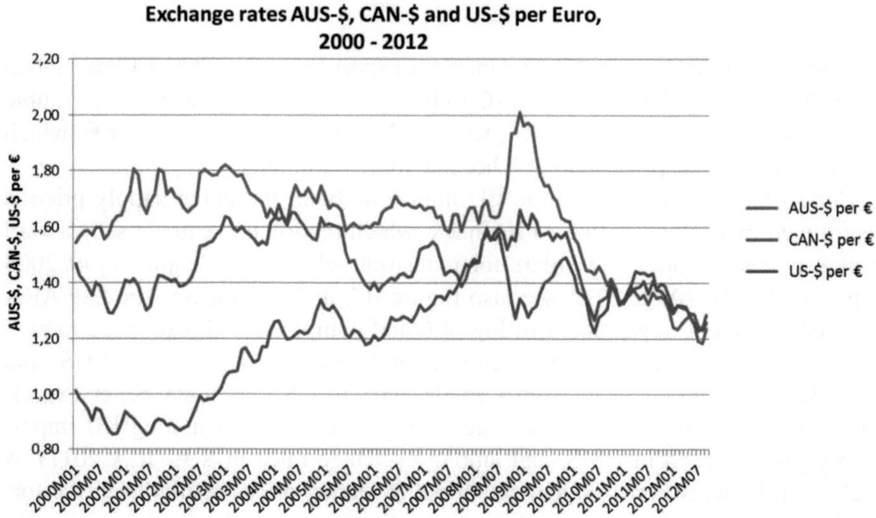

Fig. 4.4 Exchange rates of AUS-$, CAN-$ and US-$ per € from 2000 to 2012, monthly data. *Source* Eurostat (2012a)

within this period Australian exporters have faced a crucial disadvantage for trades to the EU when contracts were set in €. With a similar effect, the Canadian dollar per € declined 15 % in the same period with remarkable disadvantages for pellet

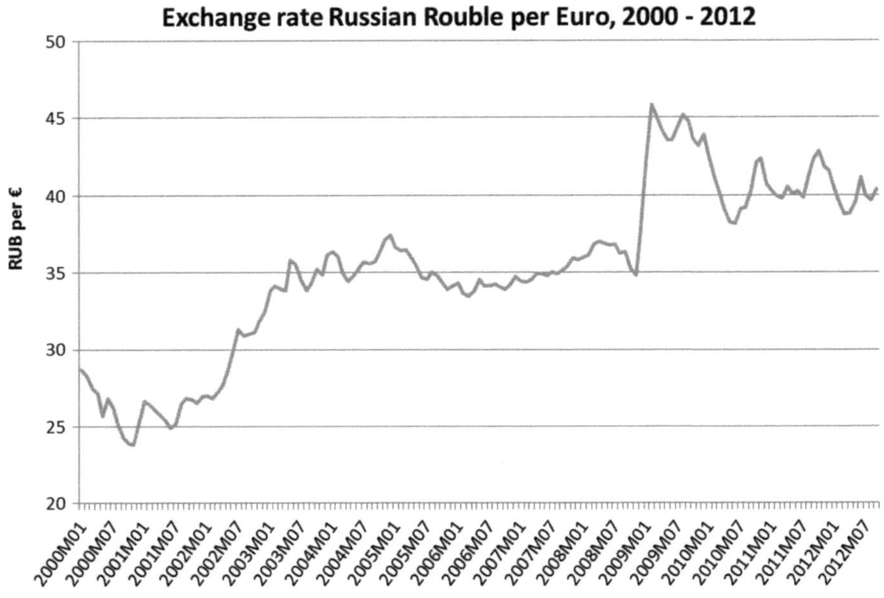

Fig. 4.5 Exchange rates of Russian rouble per Euro from 2000 to 2012, monthly data. *Source* Eurostat (2012a)

exporters. As a result, even North American export prices in CAN-$/t have fallen since 2010 due to the weak Euro (Cocchi et al. 2011). For the Russian rouble the opposite trend can be observed, with a 12 % increase of rouble per €, which means that purchase prices tied to € became more profitable.

Concerning the situation at the EU import harbour, the pellet supply price is determined by exchanges like APX Endex, which reports the actually set forward market prices in trades and orders for industrial pellets since beginning of 2008 at the ARA ports (APX 2012, see also Figure B.7 in Appendices). The CIF ARA price development is presented in Fig. 4.6 and compared with exchange rates € per US-$ and per CAN-$. There is correlation between pellet prices and US- and CAN-$, which might result from Canada and the USA as main pellet exporting countries to the EU. Besides, a general correlation exists among EU import energy prices traded in US-$, and among exchange rates in $–€ (IEA 2011). A slight shift between reported and actual pellet market prices results from the forward trade terms, which means the price is fixed one month ahead from supply (month +1).

The CIF ARA pellet price serves as reference within price negotiation between pellet exporters, traders and buyers (Interviews 2011–2013). It is significant, even if this market place has a limited amount of actors (Sikkema et al. 2011) due to few large power companies dominating the EU market. Further, it should be noticed that the CIF ARA price does not reflect price determination within longer-term contracts (Sikkema et al. 2011). As seen in Fig. 4.7, there is a certain upward

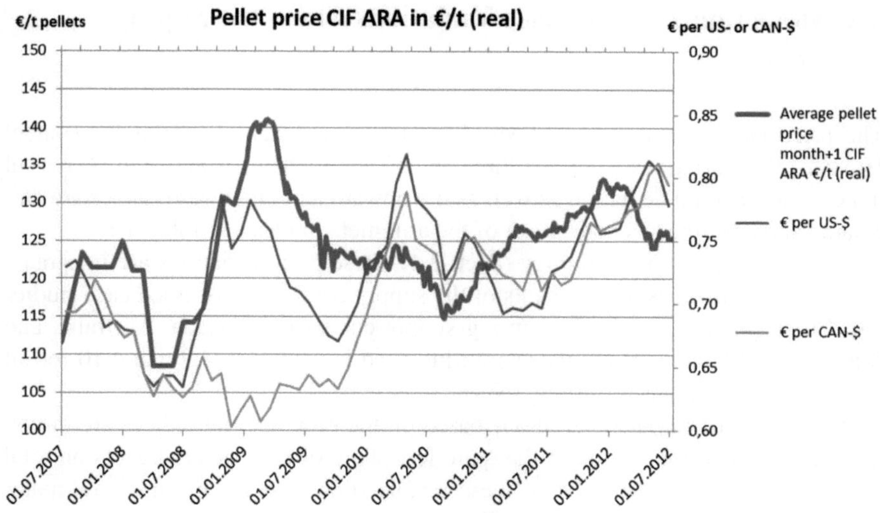

Fig. 4.6 Pellet price CIF ARA in €/t and exchange rates of US- and CAN-$ to €. *Source* APX (2012) (pellet prices up from 2008), Eurostat (2012a), WIP (2012) (pellet data 2007 for the Netherlands)

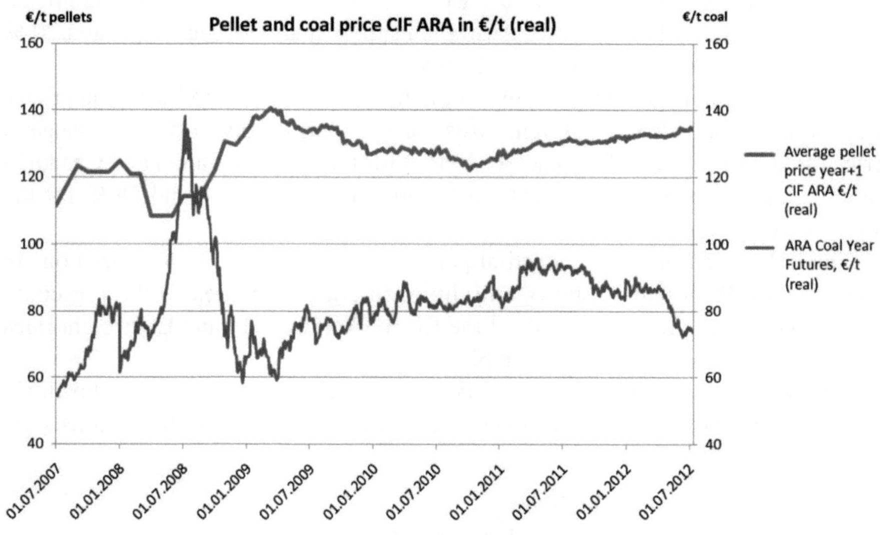

Fig. 4.7 Pellet and coal price CIF ARA in €/t. *Sources* APX (2012), EEX (2012a)

trend for both coal and pellet prices from 2010 to 2011. As found by Alakangas et al. (2012) and visible here, often a time lag of a year or more occurs between fossil fuel price hikes and increases in pellet prices because of long-term decisions to switch from fossil fuel to bioenergy.

4.2 Modelling 10-year Price Variations Along the Supply Chain

The price indices discussed in Sect. 4.1 serve as indicator for modelling latent 10-year price variations along the three pellet supply chains from Chap. 3. For all indices the standard deviation as well as the upward and downward ranges are calculated and described as percentage of the arithmetic mean, see Table 4.1.

As defined in the methodology (Sect. 2.4), these price variations are integrated into the respective cost components of the supply costs from assessed case studies (cp. Chap. 3). As a result, the varying supply costs from Canada, Australia, and Russia (free EU conversion plant) are illustrated in Figs. 4.8, 4.9 and 4.10 for all chains with natural gas for drying (NG).

As shown, the imputed variations based on the raw material indices are minor. On the one hand, this is due to the quite low share of raw material costs on total supply costs. On the other hand, these results show that the existing raw material indicators allow only little conclusion about the real cost variations for pellet feedstock (cp. Sect. 4.1.1). Thus, the issue of actual feedstock market prices will be further discussed in Sect. 4.3. Also minor influence results from the natural gas prices as marginal costs within the whole supply chain.

The exchange rate indicates only slight variations between 4 (σ) and 6 % ($R_{+/-}$) for the Canadian pellet supply to the EU. For the other cases a more significant effect from 5 (σ) to 12 % (R_{-}) for Australian supply and between 10 (σ) and 20 % (R_{-}) for the Russian supply chain can be determined.

Varying shipping freight rates turn out to have the strongest influence on prices. Depending on the share of freight costs for the total supply costs, they cause a price range of between 17 (σ) and 36 % (R_{+}) for the total Australian and Canadian supply chain or, due to minor share of freight costs, between 9 and 18 % for the total Russian supply chain.

The EU market price for industrial pellets causes 5 (σ)–14 % (R_{-}) variations in supply costs. With that, it shows only little variation compared to the preceding price risks, which are integral part of the EU market price. Inland EU freight rates turn out to have minor impact on supply costs.

The influence of fuel price variations on electricity costs is shown in Fig. 4.11. For the pellet supply prices three different kinds of upstream prices are distinguished:

- Raw material and shipping index,
- Exchange rate fluctuations and shipping index, and
- EU market price.

The standard deviation of electricity generation costs for burning 10 % pellets and 90 % coal is between 15 and 17 % of the mean costs. As for the pellet supply costs, the lower and upper ranges of co-firing costs are significantly more pronounced with 30 (R_{-})–140 % (R_{+}) of mean values. Hundred percentage coal firing (subject to coal and CO_2 price fluctuations) can cause higher fluctuations with 22–143 %

Table 4.1 Standard variation, lower and higher range limits of 10-year historic index values

Index	Standard deviation (σ) (%)	Lower range limit (R_-) (%)	Upper range limit (R_+) (%)	Sources
Wood chipping index Australia	8	88	113	ABARES (2011)
Wood product and manufacturing index Russia	9	94	105	FSSS (2012a)
Raw material index, logs and bolts, softwood, Canada	6	86	112	STATCAN (2012b)
Natural gas wholesale prices Australia	20	75	135	BREE (2012)
Natural gas consumer prices industry Russia	3	94	105	FSSS (2012b)
Natural gas consumer price index Canada	6	87	110	STATCAN (2012a)
Exchange rate AUS-$ - €	8	82	108	Eurostat (2012a)
Exchange rate RUS rouble - €	13	73	124	Eurostat (2012a)
Exchange rate CAN-$ - €	6	92	109	Eurostat (2012a)
Baltic Dry Shipping index (adjusted to €-equivalents)	48	43	201	Maersk (2010), KMI (2012)
EU pellet market price	5	86	112	APX (2012)
Freight rate index EU	0.9	99	102	DESTATIS (2012)
Hard coal EU	32	57	159	EEX (2012a), BAFA (2012)
CO_2 emission allowances	30	0.1	211	BLUENEXT (2012)

Basis 10-year time series of indices based on their annual average values (5 years Russian data, EU pellet market price and transport EU, 7 years CO_2 emission allowances), inflation adjusted with national consumer price indices from Eurostat (2012b), values are indicated in % of arithmetic mean of historic values

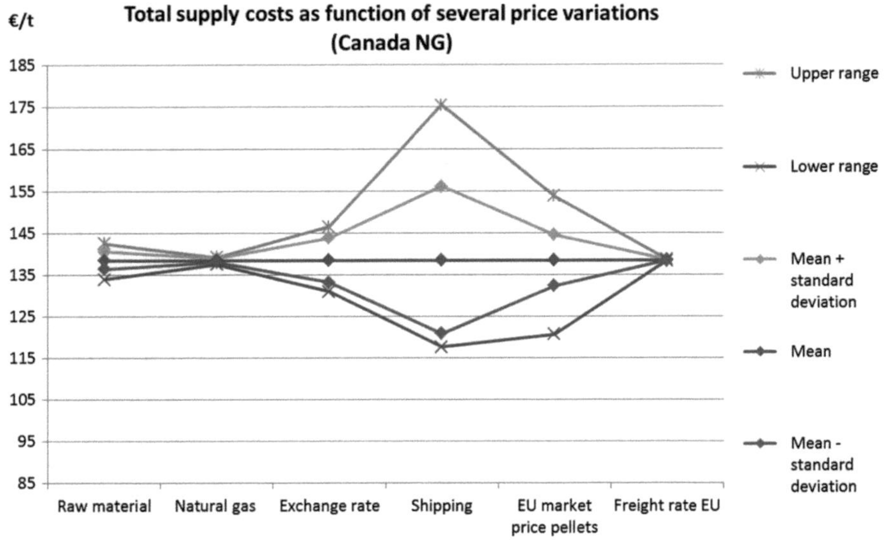

Fig. 4.8 Supply costs for Canada NG subject to selected 10-year price variations

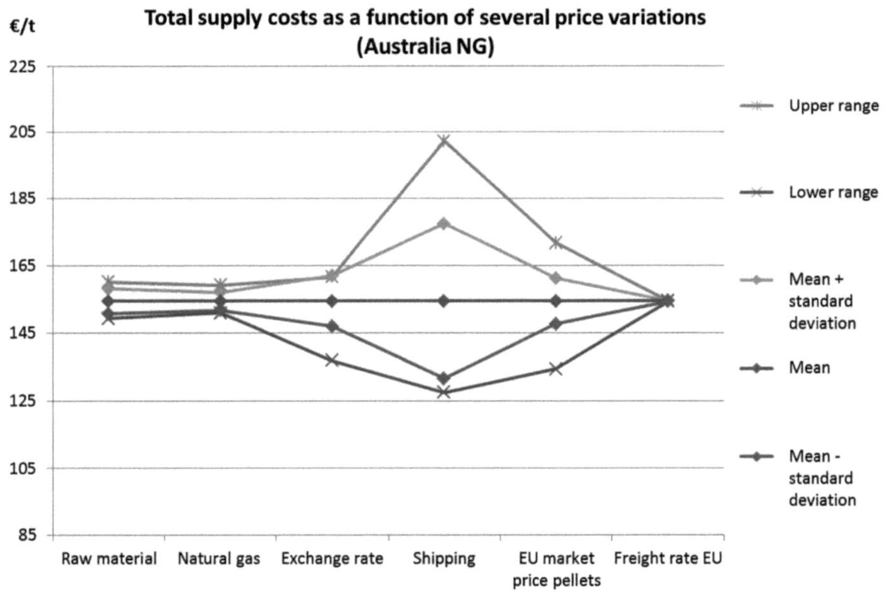

Fig. 4.9 Supply costs for Australia NG subject to selected 10-year price variations

range in the given period. Consequently, even when only 10 % of fuel input is represented by pellets, the fuel price risk can be reduced by co-firing. Beyond, the three kinds of pellet price composition differ only slightly from each other.

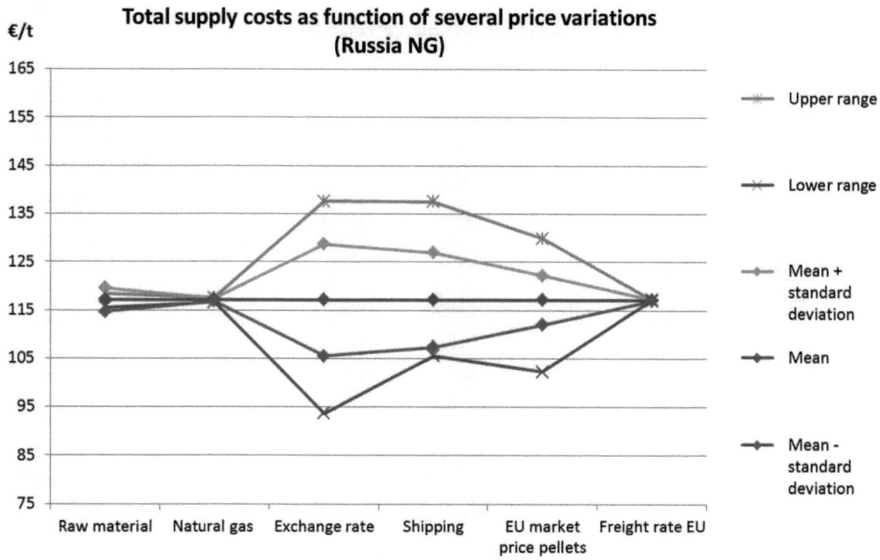

Fig. 4.10 Supply costs for Russia NG subject to selected 10-year price variations

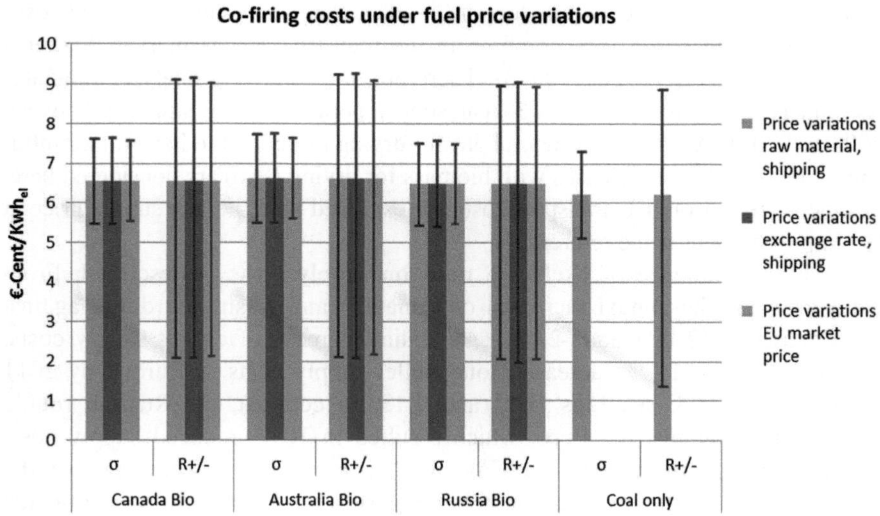

Fig. 4.11 Pellets co-firing costs (90 % coal, 10 % pellets) under fuel price variations (raw material and shipping, exchange rate and shipping, CIF ARA price)

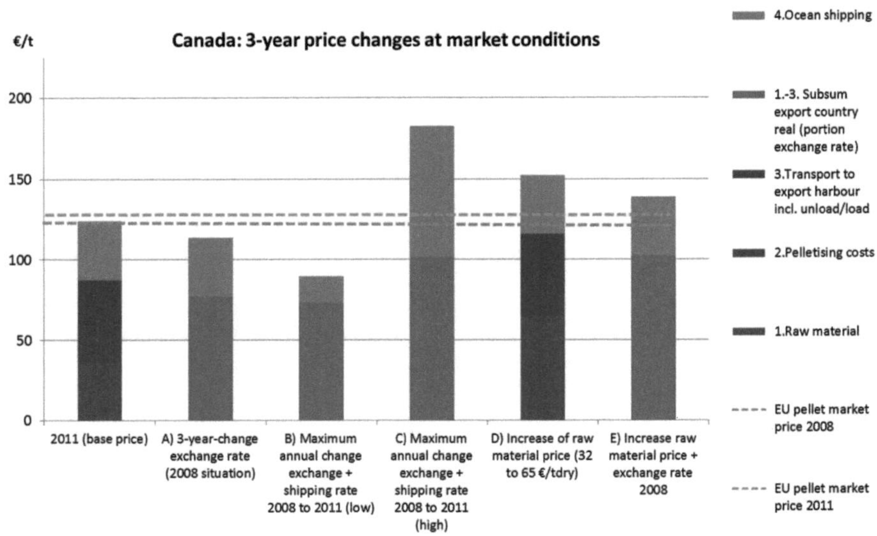

Fig. 4.12 Simulation of 3-year price changes for Canadian pellets supply

4.3 Simulation of Recent 3-year Price Fluctuations

Beside purchases on the spot or forward market, long-term supply contracts over 3 years (or longer) are very common in the industrial pellets market (Sikkema et al. 2011; Interviews 2011–2013). Therefore, the following analysis simulates relevant price changes within a 3 year supply period, i.e. the situation between 2008 and 2011. As the use of natural gas for drying turned out to have minor influence (cp. Sect. 4.2), only cases with biomass for drying (Bio) are considered here. For the same reason, EU transport costs are excluded here, i.e. the supply price at EU import harbour is the reference.

First, the influence of exchange rates on supply costs is described. From 2008 to 2011, the annual increases of Canadian and Australian dollar against Euro caused a 12 (Canada)–21 % (Australia) increase of export supply costs. This results in 8–15 % increased total pellet supply costs to Europe by 2011 compared to 2008, see Figs. 4.12 and 4.13. In contrast, the Russian rouble was devalued against the Euro, which resulted in 10 % reduced supply costs for Russian pellets in 2011, see Fig. 4.14, column B. By knowing that the Australian and Northern American pellet export to the EU was strongly affected by exchange rate fluctuations (Ferguson 2010; Grieve 2012), one can approve the evidence by Auboin and Ruta (2011) that exchange rate volatility has a significant (negative) effect on trade, especially for differentiated products like pellets. This aspect is supported by the fact that general imports from these countries to the EU dropped in the strong dollar years while Russian imports rose (Eurostat 2013a).

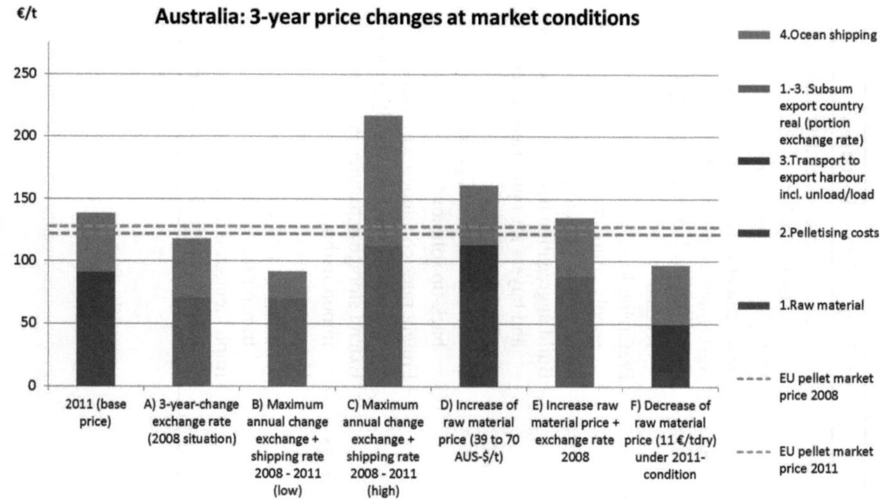

Fig. 4.13 Simulation of 3-year price changes for Australian pellet supply

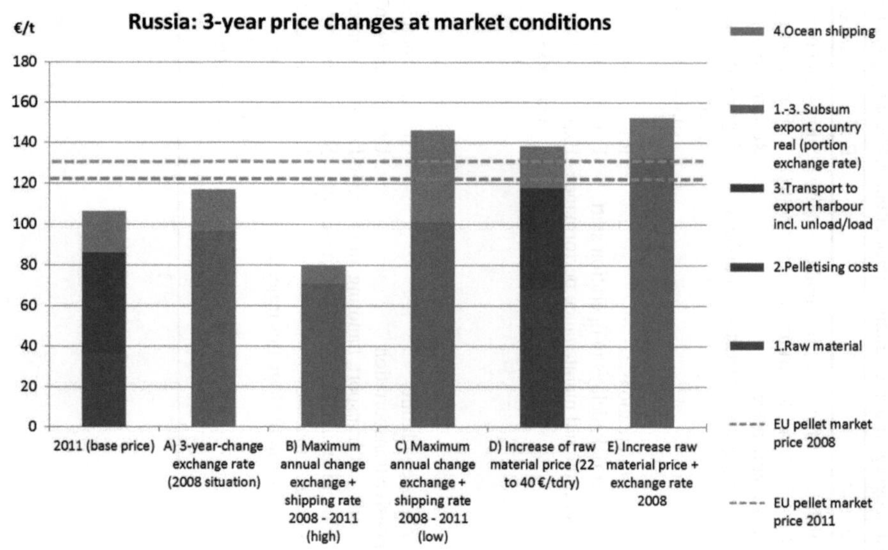

Fig. 4.14 Simulation of 3-year price changes for Russian supply

The effect of simultaneously occurring, positive or negative annual changes of both exchange and shipping rates within the 2008–2011 period resulted in a range of 66–157 % of the 2011 base price, with most pronounced ranges for Australian pellets, see columns B and C in Figs. 4.12, 4.13, 4.14. Thus, freight rates alone can easily turn supply costs to uneconomic, but also to more profitable values. Increased raw material costs (D) can effect a 16–30 % increase

Table 4.2 Price risks and related hedging strategies along international pellet supply chains

Supply chain pattern	Price risks	Relevant aspects and indicators	Hedging and de-risk strategies
Raw material	Availability of feedstock in suitable quality (just-in-)time, to certain amount, and to reasonable price Price setting among feedstock suppliers	Supply and demand for feedstock in competing industries Resource potentials Regulatory framework Monitoring of raw material prices and indexes, if available	Long-term indexed contracts with feedstock suppliers Multiple feedstock sources Involvement in resource production Agreement on price collars Controlling volumes and quality of feedstock
Pellet production	Fuel price fluctuations Investment risk, delays in operation start, operation and maintenance performance	Raw material and fossil fuel prices Financial security (equity) Experienced operation staff and sector networks	Preferring long-term and upfront supply contracts Reduction of production costs Build aggregation network of producers and buyers for optimised pellet allocation Domestic purchase of pellets as back-up solution
Transport and logistics	Freight rate volatility Expensive transportation Suitability of transport, handling and storage equipment and facilities; sufficiency of storage capacities Quality loss in the logistics trajectory	Global economic development Use of highly frequented routes Logistical capacities Seasonal access of inland waterways Logistical access to producer site & end-user Standards on durability and fines of pellets	Freight rates concluded with freight operators fixed over supply period Extend storage capacities in export and import harbours Upstream investments (storage capacity, port equipment, transportation means) Implementation of standards on biofuel handling, logistics and storage

(continued)

Table 4.2 (continued)

Supply chain pattern	Price risks	Relevant aspects and indicators	Hedging and de-risk strategies
Conversion/End-use	Prices of fossil fuel (coal) to be substituted CO_2 prices Storage capacity Price setting among few large consumers	Fuel and CO_2 price development Just-in-time delivery of fuel in certain quality and volumes EU energy and emission policy	Minor retrofit costs allow for flexible fuelling of co-firing plants thus low investment risk Access to regional biofuel as back-up Approved monitoring of market prices
Integrated aspects affecting the whole supply chain	Economic stability of export country Market price of pellets and fossil fuels Regulatory framework Evolving sustainability requirements Exchange rate fluctuations Shortfall of supplies Geopolitical risks (export taxes) Environmental risks (extreme weather events, rise or fall of water level)	Country related economic indicators Sustainability regulations, public acceptance Exchange rates Professionalising of pellet industry: experience, number, size and in time realisation of production projects and shipping activities Quality standards	Mix of long-term contracts, and of forward and spot market transactions for fuel supply Producer–buyer agreements on price collars (high-low) Vertical integration of supply chain for better control over volumes, quality and price Supply contracts in US-$ (partly minor fluctuation and no currency converting) Establishing network of producers and buyers Financial hedging of exchange rates, fuel, freight rates (forward transactions) Taking account of risk margins Standardisation of contract terms, fuel quality, sustainability

Sources Own results from previous Sections; Auboin and Ruta (2011), Bradley et al. (2010), Ehrig et al. (2011), Ehrig and Behrendt (2013), Geiver (2012), Interviews (2011–2013), Sikkema et al. (2011)

of total supply costs, whereas this rise is reduced in Canadian or completely compensated in Australian supply, when the exchange rate falls back to 2008 values (columns E). For Australia, the use of low cost sawdust could even turn the 2011 supply costs into positive values. In contrast to these huge variations, the EU pellet import price CIF ARA reflects just minor volatility with $\sigma = 5\ \%$ and R$\pm$ between 86 and 112 % of total costs within these 3 years.

4.4 Concluding Findings on Risks and Hedging Strategies

In this Section, most significant price risks, relevant aspects and respective hedging strategies—based on the previous discussion (Sects. 4.1, 4.2, 4.3) and interviews with market players—are summarised in Table 4.2. The categories follow the previously applied supply chain pattern from raw material to conversion and cross-cutting aspects. Listed aspects and de-risk measures comprise both issues already implemented and those proposed or in planning.

References

Alakangas E, Junginger M, van Dam J, Hinge J, Keränen J, Olsson O, Porsö C, Martikaninen A, Rathbauer J, Sulzbacher L, Vesterinen P, Vinterbäck J (2012) EUBIONET III—Solutions to biomass trade and market barriers. Renew Sustain Energ Rev 16(6):4277–4290

Australian Government, Department for Environment, Fishery and Forestry, Department of Agriculture, Fisheries and Forestry (ABARES) (2011) Forest and wood products statistics, Canberra. December 2011

Alderton P (2012) Personal communication with bulk parcel operator Furness Withy Australia about ocean transport freight costs for wood pellets. 23.07.2012

APX-ENDEX (2012) Historical pellets market prices. Weekly prices, Compilation by Sipke Veer, Amsterdam, 14.07.2012

Auboin M, Ruta M (2011) The relationship between exchange rates and international trade: a literature review, Staff Working Paper ERSD-2011-17, published by the World Trade Organization (WTO), Geneva, October 2011

Bundesamt für Ausfuhrkontrolle (BAFA) (2012) Drittlandskohlepreis. Durchschnittliche Einfuhrpreise frei deutsche Grenze für Kraftwerkssteinkohle. 2000–2011, available at http://www.bafa.de/bafa/de/energie/steinkohle/drittlandskohlepreis. Accessed 20 Aug 2012)

Bluenext exchange (2012) Statistics. BlueNext Spot EUA. Historical transactions from 24.06.2005 to 15.08.2012, Download of csv file at: http://www.bluenext.eu/statistics/downloads.php. Accessed 20 Aug 2012

Australian Government, Department of Resources, Energy and Tourism. Bureau of Resources and Energy Economics (BREE) (2012) Energy in Australia, Canberra, February 2012

Bradley D, Diesenreiter F, Wild M, Tromborg E (2009) World biofuel maritime shipping study. Report accomplished for IEA task 40. Ottawa, Vienna, 01.07.2009

Bradley D (2010) Canada report on bioenergy 2010, Sponsored by Canadian bioenergy association, natural resources Canada, Canadian wood fibre centre, wood pellet association of Canada, Ottawa, Sept 15, 2010

Bradley D, Hektor B, Schouwenberg PP (2010) World Bio-trade Equity Fund Study. Report accomplished for IEA task 40 bio-trade. Ottawa, April 18, 2010

Cocchi M, Nikolaisen L, Junginger M, Sheng Goh S, Heinimö J, Bradley D, Hess R, Jacobson J, Ovard LP, Thrän D, Hennig C, Deutmeyer M, Schouwenberg PP, Marchal D (2011) Global wood pellet industry market and trade study prepared for IEA bioenergy task 40: international sustainable bioenergy trade, Florence, December 2011

Statistisches Bundesamt Deutschland (DESTATIS) (2012) Erzeugerpreisindizes für Verkehrs- und Logistik-Dienstleistungen, Quartalswerte Deutschland 2006 bis 2012 data report, Wiesbaden, November 2012

European Energy Exchange (2012) ARA Coal Year Futures Request of historical prices from 2006 to 10/2012 at. http://www.eex.com/de/Marktdaten/Handelsdaten/Kohle/Coal. Accessed 15 Oct 2012)

European Energy Exchange (2012) Prices and trade volume of EU Emission Allowances, Request of historical prices and volumes at. http://www.eex.com/de/Marktdaten/Handelsdaten/Emissionsrechte. Accessed 15 Feb 2012

Ehrig R, Wörgetter M, Pointner C, Kristöfel C, Strasser C (2011) Biomass mobilisation for industrial-scale bioenergy plants. Practical approach for establishing real biomass supply pathways in Austria. In: Proceedings of the 19th European Biomass Conference and Exhibition in Berlin Eta Florence/ WIP Munich, Florence, Munich, 2011, 6–10 June 2011

Ehrig R, Behrendt F (2013) Co-firing of imported pellets—an option to efficiently save CO_2 emissions in Europe. Energ Policy 59:283–300

European Commission, Eurostat (2012a) Euro/Ecu exchange rates 2000–2011. Quarterly data, Brussels, last update 04.07.2012

European Commission, Eurostat (2012b) Harmonised consumer price index—inflation rate 1997–2011 European and Non-European countries. Brussels, 2012

European Commission, Eurostat (2013) EU27 trade since 1995 by CN8, monthly data. Brussels, last update 27.04.2012

Russian Federation, Federal State Statistics Service (2012) 25.15. Producer price indices by kinds of economic activities. Manufacture of wood and wood products. 2000, 2005–2010, data sheets, Moscow, October 2012

Russian Federation, Federal State Statistics Service (FSSS) (2012) 25.12. Indices of Producer and purchase prices for basic types of fuel-energy resources purchased by industrial organizations. Combustible natural gas at purchase price. 1995, 2000, 2005–2010, data sheets, Moscow, October 2012

Ferguson S (2010) Bioenergy development in British Columbia, Canada. mutual exchange with Europe. Presentation of the BC Bioenergy Network at 4biomass workshop, Vienna, 5th October 2010

Geiver L (2012) Contract concepts. From spot purchases to multi-year deals, producers, buyers and brokers must know their way around supply agreements, in Biomass Magazine, July 2012. pp. 25–29

Grieve O (2012) Albany biomass pellet plant winds up, ABC rural broadcast service, News release on 02.03.2012, available at: http://www.abc.net.au/rural/news/content/201202/s3422461.htm Accessed 10 Oct 2012)

International Energy Agency Ed (2011) World energy outlook 2011. Paris, 2011

Interviewed pellet producers, traders and end-users (2011–2013) Personal communication with Dusan S (2011) Black pellets manager of Vattenfall Europe AG, 08.08.2011. Hermes HD (2012) director business development biomass of Vattenfall Europe GmbH, 29.10.2012. Mertens J (2013) Biomass procurement officer at GDF Suez, 07.02.2013. Lugner M (2012) Sales manager max lugner from pellets producer Schweighofer, 15.06.2012. Pease H (2013), Senior biofuel portfolio manager for RWE supply and trading, 18.04.2013

Korea Maritime Institute (KMI) (2012) Shipping indexes. Baltic Dry Index. Daily data since 2010, available at: http://www.kmi.re.kr/Daily.do?command=List. Accessed 15 Aug 2012

Maersk M (2010) Baltic Dry Index, Compilation of daily data 1984–2010 from Bloomberg. Edited by Copenhagen Business School, Frederiksberg, 2010

Rakitova O, Ovsyanko A, Sikkema R, Junginger, M (2009) Wood pellets production and trade in Russia, Belarus & Ukraine, Market research report accomplished as subcontract for Pellets@las project, Utrecht, May 2009

Röder H (2012) Global development of bioenergy. Presentation from Pöyry consulting at University of Applied Sciences Wiener Neustadt, Wieselburg, November 2012

Senechal S, Grassi G, Nordenmark M, Vivarelli F, Junginger M, Sikkema R (2009) Logistic management of wood pellets: Data collection on transportation, storage and delivery management. Development and promotion of a transparent European Pellets Market", in frame of the task "Creation of a European real-time Pellets Atlas" for Pellets@las project, Brussels, 2009

Sikkema R, Steiner M, Junginger M, Hiegl W, Hansen MT, Faaij A (2011) The European wood pellet markets: current status and prospects for 2020. Biofuels Bioprod Biorefin 5(3):250–278

Sumetzberger H. (2012) Personal communication with the supply and logistics manager of fuel trade company Genol GmbH & Co KG. Vienna. 08.02.2012

Statistics Canada (2012a) Prices and price indexes. Consumer price index natural gas. http://www.statcan.gc.ca. Accessed 19 Jul 2012

Statistics Canada (2012b) Table 330-0007—Raw materials price indexes—2306. Monthly data, 2002 = 100. http://www.statcan.gc.ca. Accessed 19 Jul 2012

Svanberg M, Halldórsson A (2013) Supply chain configuration for biomass-to-energy: the case of torrefaction. Int J Energy Sect Manage 7(1):65–83

WIP Renewable Energies (2012) Pellets market data, request and data collection. http://www.pelletsatlas.info/cms/site.aspx?p=9138. Accessed 20 Jul 2012

Chapter 5
Summary and Discussion of Results

Abstract This chapter summarises how market aspects affect total biomass supply chain economics and how highly volatile prices influence pellet supply costs. The individual ranges of price variations in a 3- to 10-year period are outlined with raw material prices, freight rates, and exchange rates between origin currency and Euro as the predominant factors. These price aspects - compared with the final EU market price–are discussed from a holistic and close-to-market viewpoint.

Keywords Summary of price ranges • Risk drivers • Exchange rates • Risk comparison

This study demonstrates how market aspects affect total biomass supply chain economics and how highly volatile market prices influence pellet supply costs. When considering a 10-year period and individual price effects, ocean freight rates affect fluctuations between −17 to +36 % of total supply costs. Regarding exchange rates of the export country's currency to Euro, Canadian pellets costs can vary up to ±6 %, Australian pellets up to −12 % and Russian pellets up to −20 %. By now, these fluctuations are not fully represented by the EU import market price of pellets, which varies between –14 and 5 %. Other impacts as from natural gas as drying fuel during pellet production or the EU transport price indices are negligible.

Simulating a 3-year supply period between 2008 and 2011, the strongly increased dollars to Euro resulted in 12 % (Canada) to 21 % (Australia) higher export costs in € and up to 15 % higher total supply costs to the EU, whereas Russia has faced a 10 % reduction of supply costs due to stronger Euro to rouble. When annual exchange and shipping rate extremes occur simultaneously, supply costs can vary from −34 to +57 %. Also, prices for more expensive feedstock assortments can easily lead to a 16 to 30 % rise of total supply costs. Contradictorily, these actual market fluctuations are not yet reflected by the final EU pellet market price. It has to be considered that the period under observation

R. Ehrig et al., *Economics and Price Risks in International Pellet Supply Chains*, SpringerBriefs in Applied Sciences and Technology, DOI: 10.1007/978-3-319-07016-2_5,

overlaps with the recent global economic recession with strong market fluctuations in many sectors. Nevertheless, or exactly because of this, great importance should be attached to possible strong fluctuations in these global market activities.

In respect of the high individual price fluctuations in the pellet supply chain, both pellet buyers and producers have to de-risk financial exposure and must take a set of contractual provisions. A mix of longer-term and short-term contracts—agreed with several producers (buyers) from different countries—is one of the key measures to de-risk international pellet supply (see Sects. 4.1 and 4.4). Anyway, some market actors guess that extreme fluctuations, as occurred in the Australian dollar—Euro exchange rates, can hardly be hedged (Interviews 2011–2013). Even when long-term contracts exist, the risk remains that counterparties—especially smaller producers—cannot compete the new market situation after the contract has expired.

Reference

Interviews, 2011–2013: Interviewed pellet producers, traders and end-users (2011–2013), Personal communication with: Dusan S (2011) Black pellets manager of Vattenfall Europe AG, 08 Aug 2011. Hermes HD (2012) Director business development biomass of Vattenfall Europe GmbH, 29 Oct 2012. Mertens J (2013) Biomass procurement officer at GDF Suez, 07 Feb 2013. Lugner M (2012) Sales manager Max Lugner from pellets producer Schweighofer, 15 June 2012. Pease H (2013) Senior Biofuel Portfolio Manager for RWE Supply and Trading, 18 April 2013

Chapter 6
Conclusions

Abstract This chapter concludes main economic and price aspects along international pellet supply chains. These are reflected by actual framework conditions in the market and interconnections between several supply patterns and related economic impacts. Consequent securing strategies for supply agreements are provided. For example, these are fixing prices over contract period, price adjustments for inflation and exchange rate sensitive prices, financial hedging, diversifying supply and sales markets, and flexible allocation of biomass volumes via supply and demand networks. Finally, the need for further market monitoring and data research is identified to support more reliable biomass markets.

Keywords Concluding price risks • Securing strategies • Supply agreements • Fixed prices • Diversification • Market monitoring

Total pellet supply costs are mostly affected by freight rate volatility, the EU market price, exchange rate variations and actual changes in the raw material prices. The exporter's share can be affected even harder by exchange rate variations when selling in a foreign currency. Nevertheless, volatility of fossil fuel prices is much stronger and pellets can even slightly reduce it during co-firing. Also, due to easy substitution by coal and minor retrofit efforts, co-firing for electric utilities can be seen as a relatively riskless way to produce renewable electricity.

Pellet prices vary between −20 and 36 % in a 10-year-period when considering the influence of individual price risks on the final supply costs. Recent price risks occurring simultaneously in the 3-year period 2008–2011 were significantly higher with variations from −34 to +57 % of a given base price. Main causes are the exchange rate and ocean freight rates in the last few years. In contrast, the achievable EU market price for imported pellets is far less subject to fluctuations. The recent experience has shown that pellet counterparties can easily succumb this high uncertainty along with competition and fail on the market. Another result is, that 10-year price developments hardly reflect critical risks, which occur in the meantime and can determine the success or failure of a supply project.

R. Ehrig et al., *Economics and Price Risks in International Pellet Supply Chains*, SpringerBriefs in Applied Sciences and Technology, DOI: 10.1007/978-3-319-07016-2_6,

As for fossil fuels, pellet supply contracts include a number of provisions with fixed prices, inflators, specific index-based price adjustments or collars. Large market players occasionally use their access to currency, freight and fuel financial hedging. Freight rates are generally fixed between pellet buyers and vessel operators, thus the price is effectively uncoupled from the market over this period. Besides, pellet buyers follow the strategy to diversify supply contracts and use a network of producers from different economies, thus the risk of extreme exchange rate variations is reduced. Vice versa, pellet producers should diversify their sales markets. Exclusive, even price fixed supply from one producer to one end-user bears high price risks when market prices have prohibitively changed after that period.

Exporting countries with developed pellets market benefit from specialised production and logistics. This allows market actors to jointly offer larger volumes to the market, flexibly re-allocate supply and thus stabilise prices. For suppliers, also regional, neighbouring or closely linked trade markets should be considered as this avoids both expensive and volatile transportation efforts and risks connected with exchange rate fluctuations. With growth of the global pellet market a vast pool of pellet producers and buyers is expected to allow for better allocation and thus economics and stability of pellet trade.

The need for further research can be stated as follows. Deep sector-specific knowledge, setting up indices and continuous price monitoring are necessary to observe the highly complex, interconnected market and to avoid risks. Yet, the pellet market is professionalising with a growing number of experienced market actors and increasing investments in production and logistics (Interviews 2011–2013).

Thus, further investigations and monitoring in the field of feedstock prices and the pellets market, analysis of and free access to FOB prices (prices at export harbours) are desirable. The development and observation of pellet related price indices on country level could foster the confidence in newly emerging markets.

Reference

Interviewed pellet producers, traders and end-users (2011–2013), Personal communication with: Dusan S (2011) Black pellets manager of vattenfall Europe AG, 08 Aug 2011. Hermes HD (2012) Director business development biomass of vattenfall Europe GmbH, 29 Oct 2012. Mertens J (2013) Biomass procurement officer at GDF Suez, 07 Feb 2013. Lugner M (2012) Sales manager Max Lugner from pellets producer Schweighofer, 15 June 2012. Pease H (2013) Senior biofuel portfolio manager for RWE supply and trading, 18 Apr 2013

Erratum to: Economics and Price Risks in International Pellet Supply Chains

Rita Ehrig, Frank Behrendt, Manfred Wörgetter and Christoph Strasser

Erratum to: R. Ehrig et al., *Economics and Price Risks in International Pellet Supply Chains*, DOI 10.1007/978-3-319-07016-2

Below is the "Acknowledgments" part which was missed out during the initial stage of publication.

Acknowledgments

The authors would like to thank the interviewed market actors for their insightful information and efforts. Financial support in frame of the project IFSP-2 by BIOENERGY 2020+ is gratefully acknowledged.

The online version of the original book can be found under DOI 10.1007/978-3-319-07016-2

R. Ehrig (✉) · C. Strasser
Unit Resources and Technical Logistics, BIOENERGY 2020+ GmbH, Gewerbepark Haag 3, 3250 Wieselburg-Land, Austria
e-mail: rita.ehrig@gmail.com

C. Strasser
e-mail: christoph.strasser@bioenergy2020.eu

F. Behrendt
Chair Energy Process Engineering and Conversion Technologies for Renewable Energies, Berlin Institute of Technology (TU Berlin), Fasanenstr. 89, 10623 Berlin, Germany
e-mail: frank.behrendt@tu-berlin.de

M. Wörgetter
BIOENERGY 2020+ GmbH, Gewerbepark Haag 3, 3250 Wieselburg-Land, Austria
e-mail: manfred.woergetter@bioenergy2020.eu

R. Ehrig et al., *Economics and Price Risks in International Pellet Supply Chains*, SpringerBriefs in Applied Sciences and Technology, DOI: 10.1007/978-3-319-07016-2_7,

Appendix A: Key Process and Country Parameters

Table A.1 Key parameters for pellet plants

	Medium-scale pellet production	Large-scale pellet production
Annual pellet production	**40,000 t/a**	**120,000 t/a**
Pellet production rate	5 t/a	15 t/a
Annual operating hours	8,000 h	8,000 h
Total investment costs	**3,743,300 €**	**9,176,200 €** (11,286,726 € when suitable for wood chips)
Interest rate	6 %	6 %
Average service life of the pellets production facility	17.5	17.5
Raw material storage	Paved outdoor storage	Paved outdoor storage
Storage capacity[a]	1.92	1.92
Kind of storage for wet raw material	**Silo**	**Silo**
Storage capacity	0.41 %	0.41 %
Drying	**Belt dryer**	**Belt dryer (or rotary drum dryer)**
Heat demand	1,200 kWh/$t_{ev.w.}$	1,200 kWh/$t_{ev.w.}$ (belt dryer) 1,000 kWh/$t_{ev.w.}$ (rotary drum dryer)
Required electric power	140 kW	420 kW
Grinding/ sieving	**Hammermill** (additional coarse grinding unit if wood chips are considered for particle size >7 mm)	
Required electric power	110 kW	330 kW
Pellet mill	**Ring die technology** incl. driving motor for feeding of raw material and mixing screw for hot water conditioning	**Ring die technology** incl. driving motor for feeding of raw material and mixing screw for hot water conditioning
Required electric power	300 kW	900 kW
Cooling	**Counterflow cooler**	**3 Counterflow coolers**

(continued)

SpringerBriefs in Applied Sciences and Technology, DOI: 10.1007/978-3-319-07016-2,

Table A.1 (continued)

Required electric power	12 kW	36 kW
Pellet storage	Silo	Silo
Storage capacity	2.3 %	2.3 %
Peripheral equipment	Conveying systems, steel construction	Conveying systems, steel construction
Required electric power	108 kW	216 kW
Simultaneity factor for electric consumptions	0.85	0.85
Total electricity consumption	4.56 GWh/a	12.93 GWh/a (15.2 GWh/a if coarse grinding for wood chip is needed)
Efficiency of used fuel (biomass and natural gas, external boiler and related investment not considered)	90 %	90 %
Shiftwork/ Personnel		
Shifts per day	3	3
Working days per week	7	7
Persons per shift (incl. deputyship)	1.25	1.25
Personnel for marketing and administration	2 full time employees	2 full time employees

Sources Adapted from Obernberger and Thek (2010), Urbanowski (2005)
[a]Based on annual demand, annual production capacity respectively

Table A.2 Key parameter for a new coal power plant with steam turbine

Parameter	
Average nominal capacity	800 MW_{el}
Investment sum	1,040 million €
Net efficiency (without own electricity consumption)	46 %
Average operating hours	5,000 h/a
CO_2 emissions	768 g/kWh_{el} (according to EC 2007)
Deprecation time	25 a
Interest rate	6 %
Specific investment costs	1,300 €/kW
Operating costs	2 % of investment costs/a
Additional investment costs for direct co-firing of wood pellets	300 €/kW_{el} for separate feeding and grinding unit for the preparation and co-firing of pellets
Additional operation costs during co-firing of wood pellets	3 €/$kW_{el\ due}$ to increased pre-treatment efforts

Source BMU (2010), Maciejewska et al. (2006), ECN (2011)

Table A.3 Key parameters for each export country

	Canada	Australia	Russia	Sources
Hourly rate of technical staff	26.82 €/h	30.53 €/h	12.96 €/h	Berlin (2006), Raitila et al. (2009), BLS (2011)
Electricity costs	44.4 €/MWh	51.06 €/MWh	50 €/MWh	Gerasimov (2012), BC Hydro (2011), World Nuclear Association (2012)
Natural gas costs	18.26 €/MWh	18.65 €/MWh	7.33 €/MWh	Engineroom (2009), Energyshop (2012)
Biomass price (ex pellet plant)	24.73 €/tw.b. for sawdust/shavings with 36 % mc, externally purchased and transported 100 km by truck	9.06 €/t w.b. with 50 % mc, (internally) purchased by co-located saw mill, incl. handling	17.42 €/tw.b. with 55 % mc, externally purchased and transported 30 km on average	Bradley (2010), Sikkema et al. (2010), Urbanowski (2005), May (2012), Cocchi et al. (2011)

Appendix B: National and Specific Price Indices

Fig. B.1 Plantation log price index and wood chip index (basis: 1989–1990 = 100). *Source* ABARES (2011)

R. Ehrig et al., *Economics and Price Risks in International Pellet Supply Chains*, SpringerBriefs in Applied Sciences and Technology, DOI: 10.1007/978-3-319-07016-2,

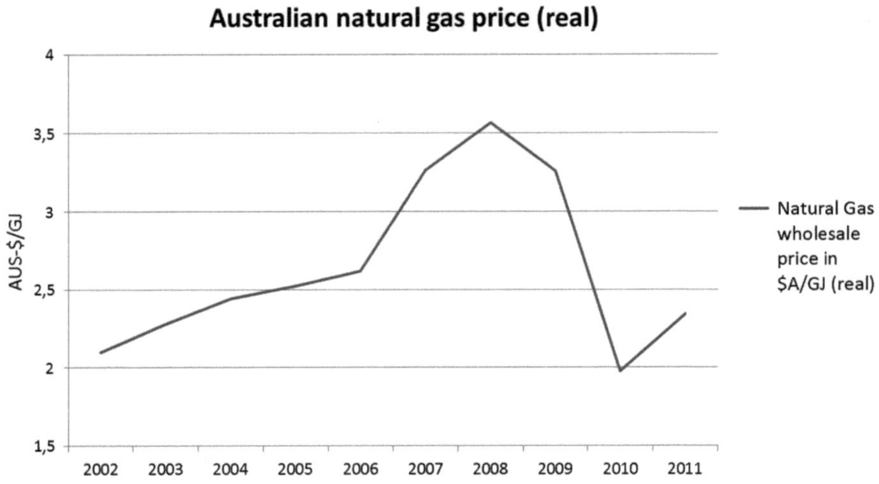

Fig. B.2 Development of natural gas wholesale prices in Australia 2002–2011. *Source* BREE (2012)

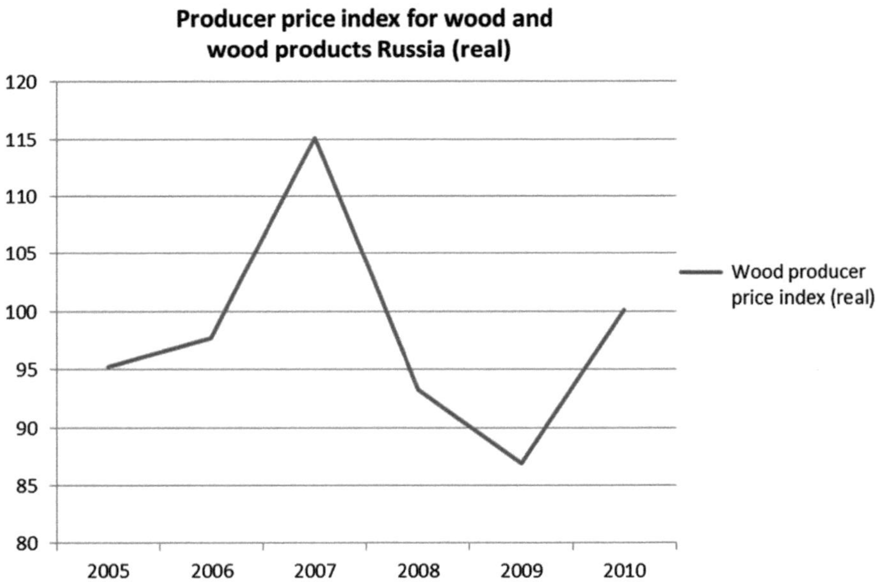

Fig. B.3 Russian wood and wood products producer price index 2005–2010. Basis: end of year percentage to end of previous year. *Source* FSSS (2012a)

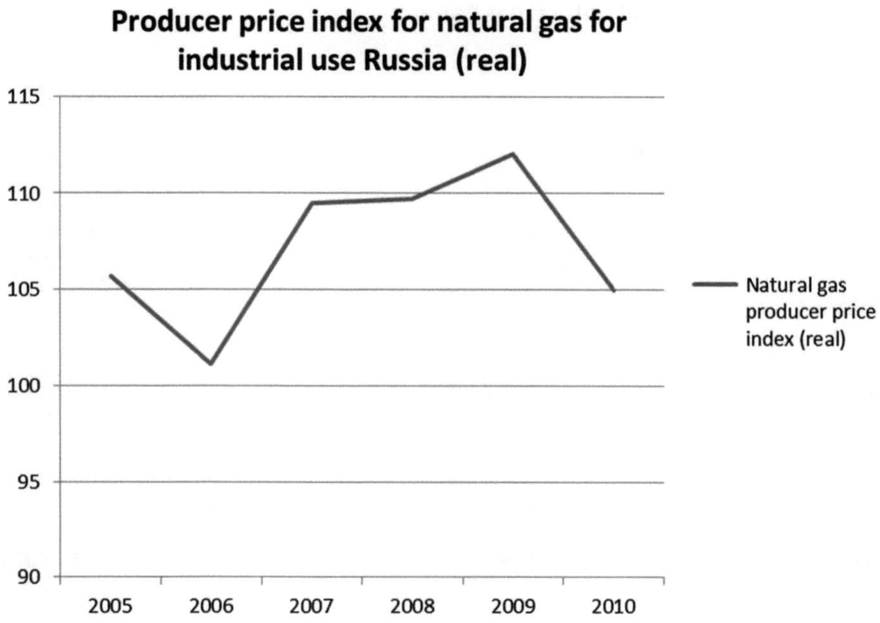

Fig. B.4 Russian producer price index for natural gas for industrial use. Basis: end of year percentage to end of previous year. *Source* FSSS (2012b)

Fig. B.5 Canadian raw material price index, logs and bolts, basis: 2002 average = 100. *Source* STATCAN (2012a)

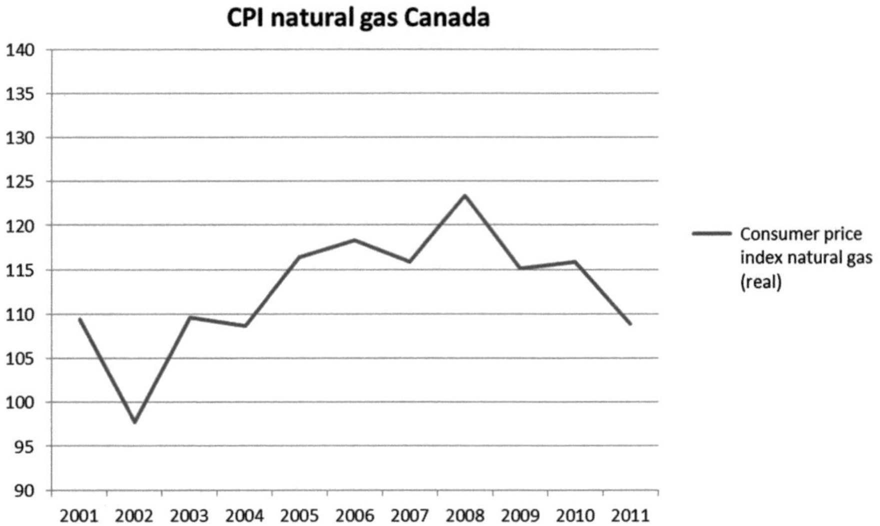

Fig. B.6 Canadian Consumer Price Index natural gas 2001–2011. Basis: 2002 = 100 (real). *Source* STATCAN (2012b)

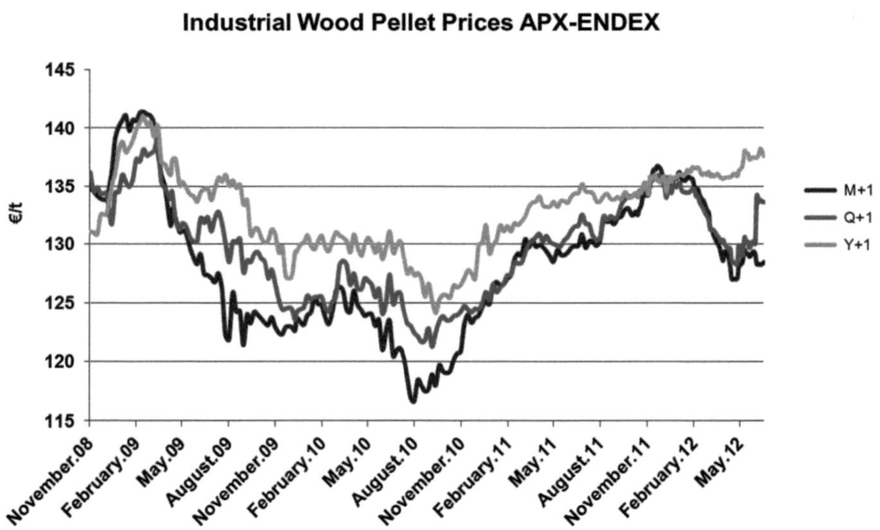

Fig. B.7 Industrial pellet prices at exchange APX-ENDEX in €/t for sales 1 month (M+1), 1 quarter (Q+1) and one year (Y+1) ahead. *Source* APX (2012)

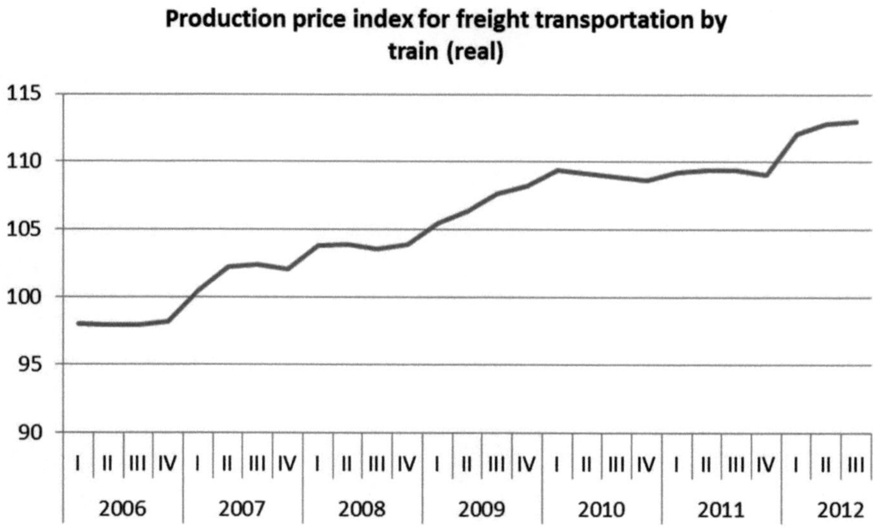

Fig. B.8 German producer price index for freight transportation by train (real). I 2006 = 100 (nominal). *Source* DESTATIS (2012)

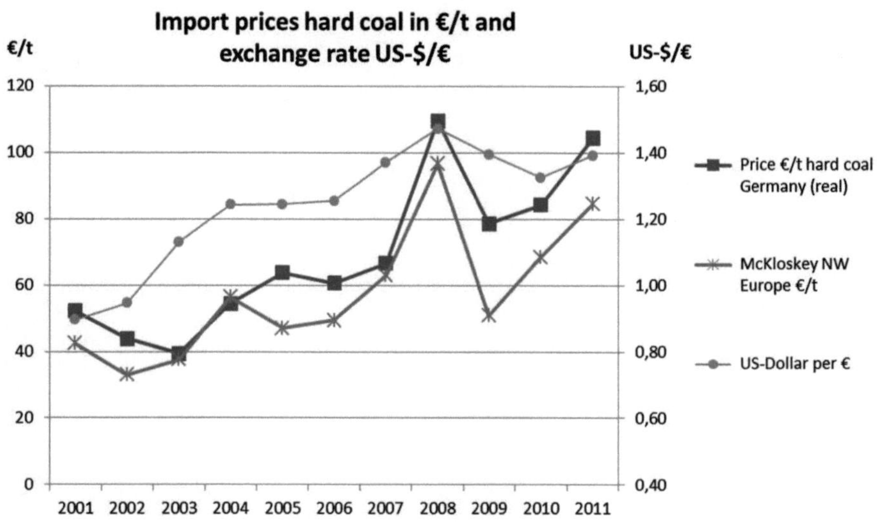

Fig. B.9 Import prices hard coal Germany and McKlosey for Northwest Europe in €/t and US-$/€. *Sources* BAFA (2012), BP (2012), Eurostat (2012a)

MIX
Papier aus verantwortungsvollen Quellen
Paper from responsible sources
FSC® C105338

If you have any concerns about our products,
you can contact us on
ProductSafety@springernature.com

In case Publisher is established outside the EU,
the EU authorized representative is:
Springer Nature Customer Service Center GmbH
Europaplatz 3, 69115 Heidelberg, Germany

Printed by Libri Plureos GmbH
in Hamburg, Germany